Electron Densities
in Molecules
and Molecular Orbitals

This is Volume 35 of
PHYSICAL CHEMISTRY
A Series of Monographs

Editor: ERNEST M. LOEBL, *Polytechnic Institute of New York*

A complete list of titles in this series appears at the end of this volume.

Electron Densities in Molecules and Molecular Orbitals

John R. Van Wazer
Ilyas Absar

Department of Chemistry
Vanderbilt University
Nashville, Tennessee

Academic Press

New York   San Francisco   London   1975

A Subsidiary of Harcourt Brace Jovanovich, Publishers

ACADEMIC PRESS, INC.
111 Fifth Avenue, New York, New York 10003

United Kingdom Edition published by
ACADEMIC PRESS, INC. (LONDON) LTD.
24/28 Oval Road, London NW1

Library of Congress Cataloging in Publication Data

Van Wazer, John R
    Electron densities in molecules and molecular
orbitals.

       (Physical chemistry, a series of monographs   ; vol. 35)
       Bibliography: p.
       Includes index.
       1.   Molecular structure.   2.   Molecular orbitals.
I.   Absar, Ilyas, joint author.   II.   Title.
III.   Series.
QD461.V32   539′.6   74-30809
ISBN 0–12–714550–8

PRINTED IN THE UNITED STATES OF AMERICA

AGAIN, YE SHALL SEE IT WITH CERTAINTY OF SIGHT!

[*The Holy Quran CII:1:7*]

Contents

Preface

Unlike physics, chemistry has attracted many persons who are not particularly adept or interested in mathematics. To these people, most of the literature dealing with molecular orbitals is distasteful, if not actually repulsive. Yet the concept of spin-paired canonical molecular orbitals obtained from self-consistent-field (SCF) calculations offers considerable insight into the electronic structure of matter, particularly molecular structures. When we first started to make plots in which the electron density in a cross-sectional plane passing through a molecule is plotted at right angles to that plane, we became enamored with the relatively high information content of this method of representing not only total electron densities but the densities of the various molecular orbitals. The Confucian maxim that "one picture is worth ten-thousand words" seems to us to be indeed true concerning cross-sectional electron-density plots. Therefore this book is primarily a picture book, with just enough text to alert the reader to some of the items he should be looking for in these plots.

In this book, we have attempted to address everyone interested in the electronic structure of molecules. We feel that the omission of all mathematics is really an advantage since this information is readily available elsewhere (e.g., see the books by Schaefer and by Pilar referenced on p. 11) and would not be of value to either the neophyte or initiate in quantum chemistry. We believe that the illustrations included in this book will be of service in explaining electronic structure to college undergraduates even at the freshman level. We have found this to be true in our own teaching, and believe that SCF molecular orbitals as depicted in this book are no more difficult for the rank beginner to understand than are the rather shopworn but basically equally valid concepts of atomic hybridization and chemical bonds now being purveyed. The molecular orbitals have the advantage of emphasizing the diffuse nature of the electrons as well as the role of this diffuseness in the bonding process.

Throughout much of Chapters 2 and 3 we have tried in words to relate each valence-shell molecular orbital to its dominant chemical-bonding contribution in order that experienced chemists (in both industry and academia) who think in terms of qualitative bonding concepts might become readily familiar with the SCF molecular orbitals and their significance within a familiar frame of reference. It seems to us that anyone who invokes atomic orbitals and their hybrids in considerations concerning chemical phenomena should at least be aware of SCF molecular orbitals other than the σ and π orbitals of diatomic and other linear molecules since the molecular orbitals result for molecular structures in exactly the same way that atomic orbitals follow from atomic structures.

This volume should also be of some value to the theoretician who is well versed in quantum chemistry. To him we commend the examples of Chapter 2 which demonstrate the effects of varying the basis set on the electron distribution within an orbital. The pictoral treatment of internal rotation at the end of Chapter 3 may also be of some interest. Our experience in dealing with a number of theoreticians is that they too sometimes have trouble in obtaining a spatial concept of electronic behavior from the usual mathematical format of their art. Spatial conceptualizing has historically been proved to be of great value in both physics and chemistry.

For those of you whose minds are spatially oriented, this book should be fun to peruse. We both hope that you will find some of the pleasure we have experienced in interrelating molecular orbitals among different molecules and among various configurations of the same molecule by means of cross-sectional electron-density plots.

Ilyas Absar
John R. Van Wazer

Full-view of an electron enlarged 10,000-fold. Note that the
electron extends beyond edge of frame in all directions.

Electron Densities
in Molecules
and Molecular Orbitals

1 Orbitals in Quantum-Chemical Calculations

A. Introduction

In the study of the electronic structure of matter, it is assumed that chemical systems such as atoms and molecules and, in turn, fundamental species such as electrons, neutrons, and protons can be represented by mathematical functions. The purpose of quantum-mechanical calculations as applied to chemistry is to find these functions, which are called the "eigenfunctions" or "wave functions" of the atom, molecule, or assemblage of atoms and/or molecules being investigated.

A theory of physical science must be able to predict as well as to explain natural phenomena and the laws that govern them. These predictive and explanative abilities are embodied in the mathematics of the Schrödinger equation [1, 2], which relates the energy of an atomic or molecular system (or their assemblages) to its wave function. Accurate wave functions may be used to calculate the energetics of chemical processes, with due allowance being made for the statistical behavior [3] of collections of atoms and/or molecules at temperatures above absolute zero. In addition, the proper arrangement of atoms necessary to make a stable molecule can be calculated without recourse in any way to experiment (i.e., on an *ab initio* basis), since the wave function corresponding to this particular arrangement will correspond to an energy minimum. Similar calculations can be used to obtain the energies associated with changes in molecular configurations [2]. Furthermore, a number of physical properties [2], such as dipole moment and diamagnetic susceptibility, may be obtained from a good *ab initio* wave function. Thus, in one sense, the long-term purpose of quantum-mechanical calculations is progressively to make experiments obsolete.

Unfortunately, the Schrödinger equation, which is a partial differential equation, cannot be solved exactly for any system containing more than one electron so that a precise analytic solution for a neutral atom can be obtained only for hydrogen. There are certain natural restrictions on these solutions which are embodied in the quantum numbers. For the one-electron problem of the hydrogen atom, there are four such quantum numbers three of which are a consequence of the three degrees of freedom in space. These three quantum numbers, which are derived from the usual solution in spherical polar coordinates, are designated n, l, and m, with n taking integer values starting with 1. For each n, there are n values of l starting with 0 and going up to $(n-1)$; and, furthermore, for each l there are $(2l+1)$ values of m starting with $-l$ and increasing in integral steps up to $+l$. The fourth quantum number, s, is introduced to account for electron spin, and s may have the values $+\frac{1}{2}$ or $-\frac{1}{2}$.

The values of $l = 0, 1, 2, 3, 4$, etc., are generally represented by the lower case letters s, p, d, f, g, etc. The various states of the hydrogen atom are designated by nl_m, e.g., $1s_0$, $2p_{-1}$, $2p_0$, $2p_{+1}$, $3d_{-2}$, $3d_{-1}$, $3d_0$, $3d_{+1}$, $3d_{+2}$. An alternative, closely related designation uses the subscripts x, y, and z for p and xy, xz, yz, z^2, and $x^2 - y^2$ for d to define a similar set of m quantum numbers with respect to the spatial geometry of the resulting functions according to the Cartesian coordinate axes x, y, and z. These various notations are simply formalities to describe the electronic states of hydrogen.

Each of the states of the hydrogen atom corresponds to a certain spatial distribution of electronic charge. This distribution may be considered

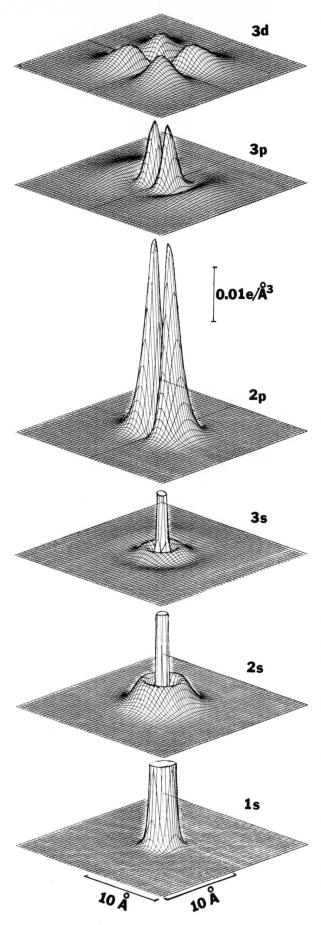

as the probability of finding the electron at any given spot or, with equal veracity, as the fraction at that position of the charge of a spatially distributed electron. This electron density is evaluated [1, 2] from the square of the wave function (or more precisely from the wave function multiplied by its complex conjugate, i.e., $\psi\psi^*$). The wave function is calibrated (by a process called "normalization") so that the summation of the electron density over all space equals unity for the single electron of the hydrogen atom.

The spatial distributions of the electron corresponding to several states of the hydrogen atom (each one of which may be called an orbital) are shown in Fig. 1.1. Each of the plots of this figure corresponds to a cut through the hydrogen atom passing through its nucleus, with the geometry of the cut being represented by the basal plane and the electron density at any point on this plane being plotted perpendicular to it. If the basal plane of these diagrams is the xy plane with the x axis running from the upper left to the lower right of each diagram, the p-type orbitals are shown as the $2p_x$ and $3p_x$ and the d-type orbital is the $3d_{xy}$. Note in the figure that the s-type orbitals are truncated in order to display them on the same scale as those of p and d symmetry. The electron density of any s orbital is a maximum at the position of the nucleus. For the 1s orbital of the hydrogen atom, the density at the nucleus is 2.148 $e/\text{Å}^3$; for the 2s it is 0.2685 $e/\text{Å}^3$, and for the 3s it is 0.0796 $e/\text{Å}^3$.

It is seen in Fig. 1.1 that, whereas the s orbitals exhibit their highest electron density at the nucleus, the p and d orbitals have no electron density at this point, but instead exhibit nodal planes (i.e., planes of zero electron density) passing through the nucleus. The differences in the energies corresponding to these various states, nl_m, of hydrogen are found to agree to good accuracy with the respective values obtained from spectroscopic measurements. The major correction is called "spin–orbit coupling," and it is necessitated by neglect of the fact that the moving electron is spinning.

B. Polyelectronic Atoms

In order to obtain meaningful solutions for polyelectronic atoms, it is necessary to use approximations in addition to those employed for

Fig. 1.1. A cross-sectional electron-density plot of the various atomic orbitals corresponding to the ground and some excited states of the hydrogen atom.

the hydrogen atom. One of these approaches, called the "Hartree–Fock approximation," involves the assumption of mutually independent one-electron wave functions that are used to build up the many-electron wave function, which can be expressed as a product of these one-electron orbitals, Moreover, since all electrons are identical, it is possible to switch any pair of them and the Pauli principle implies that the resulting many-electron wave function should be antisymmetric with respect to interchange of any two electrons. Therefore, it may be appropriately handled in the form of a determinant [2] of the one-electron wave functions, and this is called a "Slater determinant." Since it appears that a good fit to reality (i.e., a wealth of experimental data) is obtained when the one-electron wave functions used in constructing a polyelectronic atom are set up in analogy to hydrogen, this mathematical description allows the various atomic orbitals of a polyelectronic atom to be closely similar to the various states of hydrogen, so that these atomic orbitals may be described by the same set of quantum numbers.

A second common approximation, which is mathematically consistent with the previous one and which is employed in conjunction with it, is called the "self-consistent field" (SCF) approach [1, 2]. This approximation consists of a mathematical treatment in which the spatially distributed electron is considered to lie in the average potential field of all the other electrons and a series of iterations is employed to make the fields mutually consistent within the framework of the Schrödinger equation. In the above approximations, the best mathematical description of each atomic orbital leads to what is called the "limiting Hartree–Fock solution."

The energy of an atomic state obtained from a Hartree–Fock solution is never as low as the experimental energy, primarily because of the neglect of electronic correlation. Furthermore, there are relativistic effects. Although these corrections together seldom amount to more than ca. 1% of the total energy, E, of forming the atom from the isolated electrons and the isolated nucleus, the magnitude of the difference, ΔE, between the experimental and the Hartree–Fock energies is extremely large in chemical terms, especially for the heavier atoms. For the ground-state lithium atom, it has been estimated [4] that $\Delta E = 28.8$ kcal/mole, of which only 1.2% is due to relativistic effects, with $\Delta E/E = 0.61\%$; likewise, for the fluorine atom, $\Delta E = 250$ kcal/mole, of which 20.8% is relativistic, with $\Delta E/E =$

0.40%. For the sodium atom, $\Delta E = 368$ kcal/mole, of which 34.1% is relativistic, with $\Delta E/E = 0.36\%$; for the chlorine atom, $\Delta E = 1279$ kcal/mole, of which 67.3% is relativistic, with $\Delta E/E = 0.44\%$.

The correction for electron correlation accounts for details of electronic motion that are not covered by the SCF approximation (in which each electron is considered to move in the average field of the all of other electrons). In particular, this assumption of an average potential field allows electrons with antiparallel (i.e., opposite) spins to avoid each other less assiduously than is the case in reality. Various approximations [2] have been employed for estimating the contribution of electron correlation to the total energy. Such estimates generally deal only with interactions between pairs of electrons. A rough rule of thumb for atoms and ions with six or more electrons is that the correlation energy is around -50 kcal/mole per doubly filled orbital.

The relativistic correction [2] is also attributable to electron dynamics and is primarily assignable to the innermost orbitals of the larger atoms (exhibiting the larger nuclear charges). The mechanical analogy to the situation of a negative charge lying close to a large positive point charge is for the negative charge to move at an excessively high speed so that the centrifugal force will counterbalance the Coulombic attraction between the opposite charges. For electrons, this speed of revolution about the nucleus is rapid enough to lead to relativistic effects.

C. Molecular Calculations

The SCF technique can be applied to molecules in just the same way as it is applied to atoms, if we consider the molecule in the fixed-nucleus approximation [2] (which corresponds to a zero-order Born–Oppenheimer wave function). This approximation assumes that electronic motion is sufficiently faster than nuclear motion so that the nuclei may be regarded as fixed particles. Thus, the effects of small relative motions of the nuclei may be omitted from the wave function. Both theoretical and experimental evidence has amply justified the use of this approach. The functions resulting from the solution of the molecular Schrödinger wave equation are a natural extension of the atomic-orbital approach. Again, the one-electron wave functions are called "orbitals," but in this case they are molecular rather than atomic orbitals. It is important to note that except for the chemically insignificant effects introduced by the fixed-nucleus approximation, Hartree–Fock SCF molecular orbitals have exactly the same degree of significance and meaningfulness as do the Hartree–Fock SCF atomic orbitals.

For historical reasons, chemists have been at ease in thinking of atoms in terms of orbitals, even for large atoms for which wave functions could not be obtained until the last few years, when large-scale computers became generally available. In part, this touching confidence is due to the fact that the one-electron wave functions (i.e., atomic orbitals) used to describe a polyelectronic atom are similar in form to the various states (ground and excited) of the hydrogen atom for which each state is in itself a one-electron wave function. This similarity to hydrogen allowed chemists to discuss semiquantitatively the electronic structure of the heavier atoms long before it was practicable to calculate the respective wave functions.

Because there is more than one nuclear center in a molecule, the straightforward analogy to the hydrogen atom is lost and such matters as the spatial distribution of electrons in molecular orbitals have not been adequately considered until quite recently. As a result of this and because the notions of chemical bonds as electronic charge concentration entered chemistry in pre-quantum-chemical days, the molecular orbitals have seemed forbidding and cumbersome to most chemists, because it is not uncommon for the electronic charge in a single molecular orbital to be distributed around several of the nuclear centers of the molecule. Furthermore, for closed-shell systems (the usual case for stable molecules), the symmetry inherent in the arrangement of the nuclear centers carries over into the molecular orbitals which are usually described in terms of point-group notation.

Linear molecules (naturally, including all diatomics) have cylindrical symmetry and therefore may be specified by a notation which is a straightforward extension of the one used for atoms, which all have spherical symmetry. Thus, in parallel to the s, p, d, f, g, etc., atomic orbitals, the molecular orbitals of these linear structures are designated as σ, π, δ, ϕ, γ, etc. Just as the s atomic orbital has spherical symmetry, so the σ molecular orbital exhibits cylindrical symmetry. Likewise, just as each of the p, d, f, g, etc., atomic orbitals can be considered as resulting from the introduction of one, two, three, four, etc., nodal planes of symmetry to an s atomic orbital, similarly, the π, δ, ϕ, γ, etc., molecular orbitals in linear molecules exhibit one, two, three, four, etc.,

planes of symmetry, each containing the cylindrical axis of the molecule.

As a pedagogic exercise to demonstrate how molecular orbitals are formed from a combination of atomic orbitals, three-dimensional electron-density plots for any plane running through the linear axis of typical σ, π, and δ orbitals are shown in Fig. 1.2. These plots represent the electronic structures obtained by bringing two hydrogen atoms together at a distance essentially that at which maximum overlap is achieved between the participating atomic orbitals. For example, the pseudomolecular orbital marked $(\sigma_{1s})_g$ is constructed as follows: $\psi_{(\sigma_{1s})_g} = (1/\sqrt{2})(\psi_{1s}^H + \psi_{1s}^{H'})$; similarly $\psi_{(\sigma_{1s})_u} = (1/\sqrt{2})(\psi_{1s}^H - \psi_{1s}^{H'})$ and $\psi_{(\pi_{2p})_u} = (1/\sqrt{2})(\psi_{2p}^H + \psi_{2p}^{H'})$, where g stands for *gerade* (i.e., straightforward) and u for *ungerade*. Although these constructed pseudomolecular orbitals were not optimized by any SCF procedure or its equivalent, it is interesting to note that they are closely related to the symmetry states of the hydrogen molecular ion, H_2^+. For this H_2^+ ion [5], only three stable states, $(\sigma_{1s})_g$, $(\pi_{2p})_u$, and $(\sigma_{3d})_g$, were found; and these exhibit bond distances of 1.06, 4.8, and 4.2 Å, respectively. In the diagrams of Fig. 1.2 the related distance for the pseudomolecular orbitals involving the 1s atomic orbitals was also 1.06 Å, whereas a distance of 5.3 Å was employed for the pseudomolecular orbitals based on 2s or 2p atomic functions. For the pseudomolecular orbitals of Fig. 1.2 involving 3s, 3p, or 3d atomic orbitals, the interatomic distance chosen to correspond to maximum overlap was 10.6 Å.

If we consider that the plots of Fig. 1.2 approximate the probability of finding the single electron of the pseudomolecular orbital at a given position with respect to the two protons, we see that the $(\sigma_{1s})_g$ orbital corresponds to the electron lying close to the vicinity of the two nuclei and also between them. The deleted top portion of the diagram of orbital $(\sigma_{1s})_g$ in Fig. 1.2 shows individual peaks above each hydrogen nucleus with a valley between them, not unlike the uppermost part of the central portion of orbital $(\sigma_{2p})_g$. Orbitals $(\sigma_{2s})_g$ and $(\sigma_{3s})_g$ exhibit the same general form of orbital $(\sigma_{1s})_g$, except for the introduction of a spherical nodal surface around each hydrogen for $(\sigma_{2s})_g$ and two such nodal surfaces around each hydrogen for $(\sigma_{3s})_g$. These, of course, correspond to the inner nodes of the participating atomic orbitals. Note that the bonding is achieved by overlap of the outer antinodes, whereas the inner antinodes act simply as reservoirs for some of the

$(\pi_{2p})_u$

$(\pi_{2p})_g$

$(\sigma_{2p})_g$

$(\sigma_{2p})_u$

$(\sigma_{3s})_g$

$(\sigma_{3s})_u$

$(\sigma_{2s})_g$

$(\sigma_{2s})_u$

$0.01\ e/\overset{\circ}{A}{}^3$

$(\sigma_{1s})_u$

$(\sigma_{1s})_g$

10 Å   10 Å

Fig. 1.2. Electron-density diagrams of pseudomolecular orbitals generated from a linear combination of hydrogen atomic orbitals. The base plane of these diagrams passes through the two hydrogen nuclei, which have been placed so as to have maximum overlap of the atomic wave functions. (Fig. 1.2 continued p. 6.)

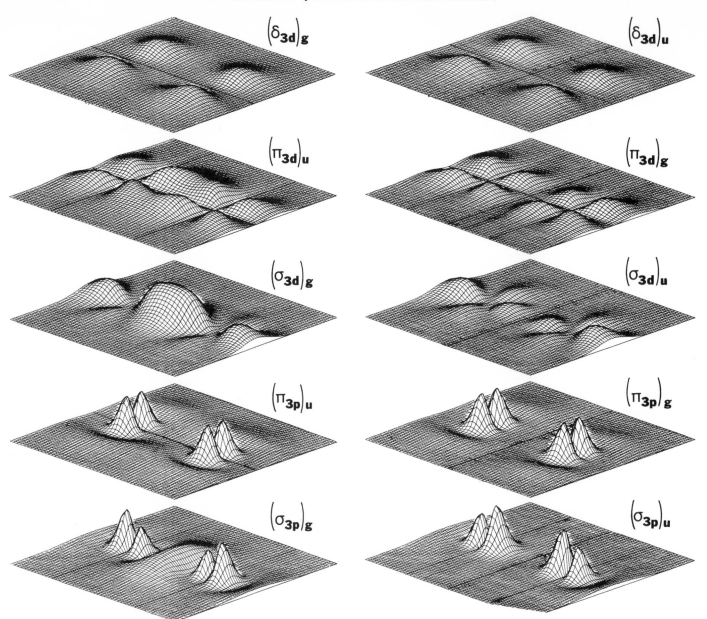

Fig. 1.2. (continued)

electron density. Orbitals $(\sigma_{1s})_u$, $(\sigma_{2s})_u$, and $(\sigma_{3s})_u$ have the same spherical nodal surfaces as orbitals $(\sigma_{1s})_g$, $(\sigma_{2s})_g$, and $(\sigma_{3s})_g$; but, in addition, each of them exhibits a nodal plane which bisects the internuclear axis. The intersection of these planes with the basal plane of the diagram is shown by the heavy line on the latter.

Pseudomolecular orbital $(\sigma_{2p})_g$ consists of the end-on interaction of a 2p lobe on one hydrogen with that on the other. Naturally, this orbital exhibits the two nodal planes corresponding to those of the participating 2p functions. However, orbital $(\sigma_{2p})_u$ has an additional nodal plane lying exactly between these two. Note that the outer

lobes of the participating 2p functions of orbital $(\sigma_{2p})_g$ and $(\sigma_{2p})_u$ act as reservoirs of electron density, with the density being transferred into the bonding region for the bonding orbital $(\sigma_{2p})_g$ and out of this region for the antibonding orbital $(\sigma_{2p})_u$. The $(\pi_{2p})_u$ pseudomolecular orbital involves the sidewise interaction of the 2p lobes on each of the hydrogen atoms so that it exhibits only one nodal plane, which passes through the internuclear axis. Likewise, orbital $(\pi_{2p})_g$ exhibits not only this plane but a plane at right angles to it, which bisects the internuclear axis. These arguments may be extended to the other orbitals shown in Fig. 1.2.

D. Basis Sets

SCF calculations may be carried out in various ways. A common procedure is to use an exponential function or a linear combination of such functions to describe each orbital, with the exponents being optimized by SCF procedures to find the set of exponents giving minimum energy. The exponential function of the form $r^{n-1}e^{-\zeta r}$, where r is the radial distance from the nucleus and ζ (zeta) is a variational parameter called the "orbital exponent," makes up the radial contribution to what is called a "Slater-type function." The exponential function of the form $r^{n-1}e^{-\alpha r^2}$, where α is another variational parameter (orbital exponent), is the radial contribution to what is called a "Gaussian-type function," with each atomic orbital being represented by a linear combination of several such functions. The various functions used to describe an atomic orbital are called "basis functions," and the number of these functions designates the size of the basis set. Although the Slater-type orbitals exhibit the same exponential form as do the analytical solutions for the various states of the single-electron hydrogen atom, it turns out that the evaluation of the various multicenter integrals involved in a molecular SCF calculation may be carried out more easily using Gaussian rather than Slater functions. At the present time, therefore, a Gaussian representation is usually preferred for *ab initio* molecular calculations.

When only one Slater-type function is used to describe each atomic orbital, the representation is called a "minimum-Slater basis set." However, even at convergence with atom optimization of the value of each orbital exponent, this basis set gives only a moderately good description of the atom. This description is considerably improved when two or more Slater-type functions are used to represent each atomic orbital. Thus, we have a minimum-Slater set as well as various extended-Slater basis sets, such as a double-zeta, triple-zeta, etc., set corresponding to the use of two, three, etc., Slater functions to describe each atomic orbital. If, say, a different number of Slater functions is employed to describe the various orbitals (e.g., a double zeta for a 1s and a triple zeta for a 2s), the overall description is simply included in the generic class of extended-Slater basis sets. Since the exponential form of Gaussian functions is different from that of analytical solutions for the hydrogen atom, two to four Gaussian orbitals are required as replacements

for each Slater orbital to get about the same total energy of the atom or molecule. Quite large Gaussian basis sets are generally employed in modern calculations, with optimization of all of the exponents in the respective atoms, as well as the usual optimization of the coefficients that weight the contributions of each Gaussian function in the linear combination employed. It is the exponent that determines the "orbital radius" for each atomic orbital, with a larger exponent corresponding to a smaller radius. If several exponents are employed to describe a given atomic orbital, the orbital radius then results from a weighting by the respective coefficients of the contributing exponential functions.

In dealing with Gaussian basis sets, it is common practice to employ the full set of exponents of a given symmetry (i.e., l quantum number) for each atomic orbital of that symmetry (i.e., the same set of exponents are used for the 1s, 2s, and 3s orbitals of a given third-period atom). Under these conditions, the members of an atom-optimized set of Gaussian exponents used for each symmetry are spaced approximately evenly on a log scale, with some bunching up of the larger exponents because they predominate in the description of the higher-energy inner orbitals and because the optimization is carried out through minimization of the total atomic energy. Furthermore, it is found [6] that the values of the atom-optimized exponents obtained for a selected number of Gaussian functions of a given symmetry are practically unaffected by the number of Gaussian functions employed to describe the orbitals of different symmetry. Thus, for carbon, essentially the same individual values were obtained for a set of, say, nine s-type functions when optimized with anywhere from one to six p-type functions.

Of course, calculations on molecules may be carried out just as they are on atoms, with full optimization of all coefficients and exponents of appropriately chosen exponential functions. However, it is found to be computationally expeditious to use a linear combination of atomic orbitals (LCAO) in the SCF calculation [1, 2]. Just as an atomic orbital can be described in a Gaussian or a multiple-zeta Slater basis set in terms of a linear combination of orbital functions, a molecular orbital may be treated as a linear combination of atomic orbitals. These, in turn, are often linear combinations themselves. Since the chemical bonding in molecules really represents only a small perturbation of the constituent atoms, it is quite common to optimize the exponents of the filled atomic orbitals in the ground-state atom and to use these "atom-optimized" exponents in the molecular calculation. This may lead to some error in the description of an atom within a molecule when a minimum-Slater basis set is employed. However, when several atom-optimized exponents are used for each atomic orbital without contraction (see below), the molecular optimization of the coefficients of the individual functions used to describe a given atomic orbital can make allowance for an effective change in radius of this orbital when going from the atom to the molecule in question. This method of indirectly achieving molecular optimization of atomic-orbital radii is particularly effective when the entire range of Gaussian exponents for a given orbital symmetry contributes to each individual atomic orbital of that symmetry.

Two common ways of lowering the cost of an LCAO–MO–SCF calculation are the use of orbital contraction and symmetry adaption. In orbital contraction, the original linear combination of basis functions used to describe each atom in a molecule is reduced by bunching several basis functions together through the use of linear combinations, with the integrals being evaluated for the overall basis set and the SCF optimization being carried out on the contracted set. This procedure has the disadvantage of reducing the capability of adjusting atomic-orbital radii to fit the requirements of various molecules, as described for uncontracted orbitals above. Symmetry adaption merely means that the individual functions of a complete basis set (or the combinations of these functions) which are unneeded for a particular molecular symmetry are deleted so that computer time is not wasted in calculating large numbers of zero-valued integrals and the full Hartree–Fock matrix may be replaced by smaller matrices, one for each pertinent symmetry type, to reduce the cost of matrix diagonalization.

When the SCF procedure is used for energy optimization, the total energy corresponding to the mathematical description afforded by a given basis may be obtained to a high degree of accuracy. Thus, for a given choice of say, Slater or Gaussian exponents, the final SCF energy after a sufficient number of iterations will always be the same, and this basis-set-dependent value is called the "convergence limit." However, if increasingly larger basis sets are used, it is found that the convergence limit approaches a limiting value of the total energy and this limiting energy, corresponding to an infinitely large basis set, is called the "energy

at the Hartree–Fock or SCF limit" [2]. This Hartree–Fock energy is always the same for any given atom or molecule and is independent of the choice of the basis set, as well as of the type of function used to describe the orbitals making up the basis set. In the study of molecules, various odd combinations [2] of orbitals have been tried in order to minimize calculational expenses. From these studies it is clear that the LCAO approximation may be extended in peculiar ways. Thus, the Hartree–Fock limit might possibly be obtained by using only s-type orbitals, which are scattered around in space and not necessarily centered on any nuclei. In the extreme, it may well be that one can use an infinitely large basis set consisting of practically any or many kinds of orbital functions expressed in any convenient mathematical form scattered around or located at an arbitrary or optimized point in space. Although the application of this approach to atoms has not received much attention, there is no basic reason that it cannot be generally employed. If these ideas seem hard to visualize, note that a properly chosen s orbital (described by, say, several Gaussian functions), when centered at the appropriate distance from the selected nucleus along the positive x axis with another of opposite sign similarly located along the negative x direction, gives a good description of a p_x orbital. (This particular type of representation is sometimes called a "Gaussian lobe function" [2].)

A commonly used formality in chemistry is the discussion of molecules and molecular orbitals in terms of the atomic orbitals making up the LCAO basis set. If the atomic orbitals are properly chosen (as in a minimum-Slater basis set) and are each centered on their respective atoms, the formality of discussing the molecule in terms of atomic orbitals can be useful and edifying. However, in such discussions it is important that the basis set be "balanced" [7] so that the choice of the atomic orbitals does not lead to bookkeeping errors in the atomic charges, as analyzed either per atom or per atomic orbital. Except for the case of the minimum-Slater basis set, it is difficult to judge when a combination of atomic orbitals is well balanced and the literature is full of examples of extremely poorly balanced basis sets. When too many atomic orbitals are assigned to one atom as compared to one or more of the others in a limited basis set, these atomic orbitals may effectively represent the starved atom(s), although being formally assigned to the other atom. An extreme case of this situation is found for the one-center calculations in which all of the atomic orbitals are centered on a single nucleus so that, even at the Hartree–Fock limit, this atom would bear the entire charge according to the LCAO bookkeeping. Problems in balancing basis functions, particularly with Gaussian-type functions, may be found even within a given atom when, say, too many p-type functions are employed in comparison with the number of s-type functions used in the description of the atom or molecule. Under such circumstances, the combination of p-type orbitals may flesh out a starved s orbital, but such substitution may lead to a distorted representation of reality.

In computations on atoms, inclusion of normally unfilled atomic orbitals (such as those of larger l quantum number) has no effect on the energy, since optimization leads to zero coefficients for them. For molecules, however, the situation is quite different in that the Hartree–Fock limit cannot be achieved in a description in which the atomic orbitals are centered on the nuclei without invoking orbitals of higher l quantum number—atomic orbitals that would be completely unfilled in the isolated constituent atoms. In other words, the convergence limit using only s- and p-type functions for a molecule involving atoms of the second and third period (e.g., C or Si, respectively) will always be less stable than the Hartree–Fock limit, even though an infinite basis set of s and p orbitals (centered on the atoms) is used. For example, a reasonable approximation to the Hartree–Fock limit is not approached for the water molecule [2] until at least one fivefold set of d-type functions is allowed to the oxygen, with perhaps a set of p orbitals to each hydrogen and even a set of f orbitals to the oxygen. The exponents of these higher orbitals must ultimately be obtained from optimization in the molecule, since the atomic-orbital radii obtained from SCF calculations on the appropriate excited state of the atom turn out to be several times larger than the radii resulting from molecular optimization. (Note that atomic radii are inversely related to the exponents.) This difference is attributable to the fact that the higher-symmetry orbitals used in molecular calculations must exhibit good overlap with the other valence-shell atomic orbitals and, to do this, must exhibit about the same atomic radii. The effect of these higher atomic orbitals is primarily to distort the shape of the molecular orbitals obtained without their use, thereby allowing the charge of each molecular orbital to achieve its optimum spatial

distribution within the average field of all of the other electrons in the system.

It now appears that a fairly good approximation to the Hartree–Fock limit for a molecule can be obtained by using an atom-optimized double-zeta Slater basis or its Gaussian equivalent plus the addition, in either case, of one molecularly optimized atomic-orbital manifold on each atom, with these additional (molecularly optimized) orbitals exhibiting the next symmetry quantum number (e.g., d as compared to p) beyond that of the outermost filled one-electron orbital of the free atom.

E. Semiempirical Calculations

Until recently, the majority of quantum-chemical calculations on molecules were carried out with what are called semiempirical methods [8, 9], such as the Hückel-type, CNDO, INDO, MINDO, and PPP procedures. In all of these the inner orbitals are neglected by treating them and the nuclei as a point-charge core. In PPP and many of the Hückel-type calculations, only the π-molecular orbitals are considered [1, 8, 9]. Since semiempirical methods were developed before extensive calculations had been done on atoms, it is common to obtain the valence-orbital exponents from a set of rules (Slater's rules [1, 8]). Generally, the approximation of zero differential overlap is employed. According to this approximation, those Coulombic integrals corresponding to an electron being associated with two different centers are completely neglected. The remaining integrals are then lumped into a few parameters, some of which are estimated or adjusted by use of atomic spectral data, with the additional parameters being set to give reasonable values for a specified calculated property.

Although in principle a semiempirical calculation would not be expected to lead to as good a description of a given physical property as that obtained when the same minimum-Slater basis set was used in a computation in which all of the integrals were evaluated (i.e., an *ab initio* calculation), the semiempirical results are usually better for the property for which the parameters were optimized. This is an example of the use of semiempirical methods as extrapolatory devices that transcend their quantum-mechanical significance. Sometimes the results are even better than those corresponding closely to the Hartree–Fock limit, using very large basis sets. In these cases, the favorable results are surely attributable to inadvertent beneficial error in the approximations rather than to a subtle inclusion of electron-correlation corrections.

F. Electronic Ionizations and Transitions

The energies of the one-electron wave functions (i.e., atomic or molecular orbitals, respectively) in polyelectronic atoms or in molecules are only indirectly related to the experimental ionization energies of their respective electrons. In the case of hydrogen, the removal of an electron from the atom leaves the bare nucleus; but, in the case of polyelectronic atoms or molecules, the removal of an electron leaves a hole for which the other electrons then have to adjust. Thus, if an electron in a particular atomic or molecular orbital is deleted from an SCF wave function, the average field in which each of the other electrons exist is disturbed so that it is necessary to go through the entire SCF procedure to minimize the energy for the hole state. In spite of this, and also in spite of the absence of corrections for the various approximations used in obtaining the particular atomic- or molecular-orbital energy, orbital energies are found to afford a reasonable approximation to experimental electronic binding energies. Indeed, Koopmans has shown theoretically that, for a closed-shell molecule, the orbital energy computed in an *ab initio* SCF calculation is approximately equal [10] to the ionization potential of the electron of that orbital. In the open-shell case, the orbital energies do not have exactly the same meaning as in the case of closed shells because of the presence of some small off-diagonal terms—a situation that implies that these orbital energies do not correspond exactly to Hartree–Fock eigenvalues. Furthermore, there are complications introduced because degenerate open shells cannot be described as single Slater-determinantal functions. However, as these effects generally cause only small perturbations, Koopmans' theorem is often applied also to open-shell atoms and molecules. For the closed-shell systems, which correspond to the usual ground-state molecules, Koopmans' theorem is equivalent to assuming frozen orbitals and vertical ionizations (i.e., exactly the same nuclear arrangements in the molecule as in its ion resulting from removal of the electron).

In the LCAO–MO–SCF procedure, the mathematics requires that there be as many molecular orbitals (MO's) as the number of atomic-orbital functions (AO's) put into the calculation. This means that the results include a specified number of unfilled molecular orbitals, called "virtual orbitals," each of which is normally found to exhibit a positive rather than the negative energy of a filled orbital. Unfortunately, electronic transitions to or between excited states cannot be estimated for any practical use by taking the difference between orbital energies. However, practitioners of semiempirical theory sometimes employ the virtual orbitals in electronic transition-energy calculations with a varying degree of success. In such cases, any physical significance ascribed to the virtual orbitals must be attributable to the semiempirical calculation acting more as an extrapolation device than as an application of electronic theory to the study at hand.

G. Molecular-Orbital Charge Distributions

Since SCF molecular orbitals have exactly the same significance for polynuclear systems that SCF atomic orbitals have for polyelectronic atoms, and since the energies calculated for these orbitals are acceptable approximations to the respective physically observed electronic binding energies, it seems reasonable that chemists should become more familiar with these molecular orbitals. To this end, the SCF molecular orbitals for a number of common molecules are presented in the remainder of this book. Just as the spatial charge distributions of atomic orbitals have proved so useful in the comprehension of chemistry in the past, so should familiarity with the charge distributions of molecular orbitals lead to better understanding in the future. To this end, many three-dimensional electron-density plots are presented in the following chapters. As in the case of Figs. 1.1 and 1.2, which have already been presented, these plots depict the intensity of the electron density perpendicular to the geometrical plane in which it is measured.

Most of the wave functions for which electron-density plots and other properties are presented result from the use of a Gaussian basis set sized to give about the same total energy as a minimum-Slater basis set. In these calculations, therefore, the hydrogen atoms are each described by the use of two or three s-type atom-optimized Gaussian exponents. Atoms of the second period (e.g., C, N, and O) are generally each represented through the use of five s-type and two p-type atom-optimized Gaussian exponents, whereas for the elements of the third period (Si, P, and S), nine s-type and five p-type atom-optimized Gaussian exponents are employed, with a single molecularly optimized exponent being used to describe the d orbitals when they are included. The major error in this restricted-basis-set Gaussian representation of the molecules is found in the fact that the three-dimensional electron-density plots exhibit rounded rather than pointed maxima at the positions of the hydrogen nuclei. This matter is discussed in more detail in Chapter 2.

H. Orbital Transformations and Calculations Not Involving Orbitals

Methods are available for transforming [2] one set of orbitals into an entirely equivalent but different set. This technique has been employed to convert the delocalized SCF molecular orbitals (of the type discussed in this book) to localized versions that correspond more or less to the average chemist's notions of chemical bonding. Of course, the resulting energy-expectation values of these versions have no particular physical significance and the particular choice of the set of delocalized orbitals is rather arbitrary.

Even beyond the Hartree–Fock limit, there are, of course, additional contributions to the energy resulting from relativistic and electron-correlation effects. Even though each of these contributions involves energies often very much larger than those found for chemical reactions, there is considerable cancellation because transfer of a given atom from one molecular environment to another represents only a relatively minor perturbation of that atom. It therefore appears that relativistic effects are generally inappreciable in a quantum-mechanical calculation of the energy of a chemical reaction. The contribution of electron correlation to such reaction energies often amounts to only about a third of the total value of ΔE or ΔH. There are some unusual cases, however. For example, even at the Hartree–Fock limit, SCF calculations show the fluorine molecule, F_2, to be unstable with respect to its atoms. Estimations of the molecular extracorrelation energy involved in the reaction whereby two fluorine atoms form a molecule demonstrate that stabilization of this particular molecular structure is achieved through electron correlation.

There are two generally accepted methods for obtaining solutions of the Schrödinger equation that are better than the SCF solutions at the Hartree–Fock limit in that they include electron correlation. The most common of these methods is called "configuration interaction" and involves linear combinations of the appropriate excited SCF states (constructed from the virtual orbitals) with the SCF ground state. In order for the configuration-interaction method to be highly accurate, it is necessary that an infinite number of such linear combinations be approached.

The alternative attack, which has not received much application to molecules, consists of the many-body perturbation methods for solving the Schrödinger equation. This also starts from an SCF calculation near the Hartree–Fock limit. An important result stemming from either configuration-interaction or many-body-perturbation methods is that the individual orbitals (either atomic or molecular) are lost during the mathematical operations. Of course, these methods are suitable, at least in principle, for calculating energy differences between various states, such as the states existing before or after the expulsion of a particular electron. However, values corresponding to orbital energies and/or spatial distributions of the electronic charge of an orbital cannot be extracted from these kinds of wave functions. Thus, we see that the concept of an orbital is intimately related to the particular mathematical approach in which each individual electron of a polyelectronic atom or molecule is considered to be moving in the average field of the other electrons.

It is true that a given set of molecular orbitals may be transformed into any number of mathematically equivalent sets; the same thing may be said of atomic orbitals. Therefore if atoms are to be discussed in terms of s, p, d, etc. atomic orbitals, then the exactly equivalent discussion for molecules must be founded on the SCF molecular orbitals, the electron densities of which are the subject matter of this book.

REFERENCES

1. F. L. Pilar, "Elementary Quantum Chemistry." McGraw-Hill, New York, 1968.
2. H. F. Schaefer, III, "The Electronic Structure of Atoms and Molecules." Addison-Wesley, Reading, Massachusetts, 1972.
3. N. Davidson, "Statistical Mechanics." McGraw-Hill, New York, 1962.
4. A. Veillard and E. Clementi, *J. Chem. Phys.* **49**, 2415 (1968).
5. D. R. Bates, K. Ledsham, and A. L. Stewart, *Proc. Roy. Soc., Ser. A* **246**, 215 (1953).
6. C. J. Hornbach, "Optimum Atomic Gaussian Functions for Atomic and Molecular LCAO-SCF Calculations," Ph.D. thesis. Case Institute of Technology, Cleveland, Ohio, 1967.
7. R. S. Mulliken, *J. Chem. Phys.* **36**, 3428 (1962).
8. J. A. Pople and D. L. Beveridge, "Approximate Molecular Orbital Theory." McGraw-Hill, New York, 1970.
9. M. J. S. Dewar, "The Molecular Orbital Theory of Organic Chemistry." McGraw-Hill, New York, 1969.
10. W. G. Richards, *Int. J. Mass Spectrom. Ion Phys.* **2**, 419 (1969).

2 Electron Densities and Shapes in Atoms and Molecules

A. Some Typical Atoms

When a 2p orbital of an atom is plotted so that the value of its wave function, ψ, is measured on an axis vertical to a coordinate plane passing through the atomic nucleus, the resulting graph exhibits a hump on one side and a hollow on the other side of the nodal plane of the orbital. This is illustrated for the oxygen atom by the top diagram of Fig. 2.1, for which the wave function was calculated using five s-type and two p-type Gaussian functions, i.e., a (52) basis set. The second diagram in this figure corresponds to the electron density, $\psi\psi^*$, which, being the square of the wave function, exhibits sharper peaks than does the wave function itself, with both peaks being positive. The three-dimensional shape of the electron-density distribution may be demonstrated by showing the surface corresponding to some arbitrarily chosen electron density. Such an orbital-shape plot is shown at the bottom of Fig. 2.1 for this oxygen 2p orbital, using an electron density of 0.14 $e/\text{Å}^3$.

Figure 2.2 presents the total electron density of the oxygen atom in its ground state, ^3P, as calculated using a minimum-Slater basis set [1] (min. Sl.), a moderately sized Gaussian basis set [2] employing five s and two p exponents

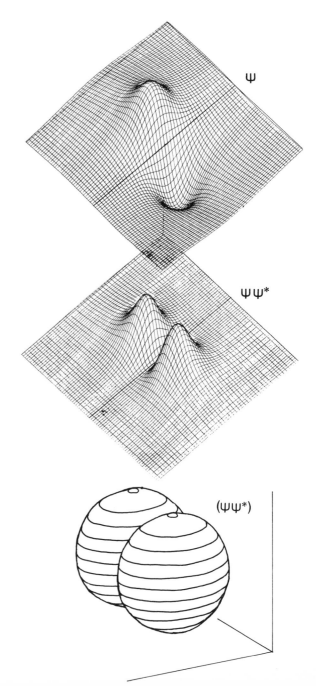

Fig. 2.1. Various representations of a 2p orbital of oxygen as calculated from a (52) Gaussian basis set (i.e., involving five s-type and two p-type functions). The plot labeled ψ shows the wave function as measured perpendicular to a geometrical plane in the center of which lies the oxygen nucleus; $\psi\psi^*$ is the respective electron density, with this density being plotted perpendicular to the basal plane; $(\psi\psi^*)$ gives the shape of the oxygen 2p orbital by showing the three-dimensional surface at which the electron density equals 0.14 $e/\text{Å}^3$.

(52), a double-zeta Slater basis set (2ζ) [3], and
an even larger (exten. Sl.) Slater basis set [4]
lying close to the Hartree–Fock limit. The plots
of this figure are scaled so that the electron-density
distortions of atoms due to chemical bonding
would show up well if they were present. In order
to conserve space, the pointed tops of these plots
are truncated. However, the cutoff sections of
these plots look alike, although they do not ter-
minate at quite the same maximum electron
density.

Figure 2.2 shows that for the oxygen atom there
is little difference in the total electron density as
described in a minimum-Slater or moderately
sized Gaussian basis set or in a near-Hartree–Fock
Slater basis set. Similar studies on the individual
atomic orbitals show that the form of their electron-density distributions is also independent to
about the same amount as is the total electron-
density distribution with respect to the choice of
the mathematical description. If the mathematical
description is about as good as a minimum-Slater,
natural, or hydrogenic orbital, the electron-density plots for all of the atoms, except hydrogen,
turn out to be superficially indistinguishable from
those obtained at the Hartree–Fock limit.

Hydrogen in its ground state represents a
special case in that it can be derived analytically.
The minimum-Slater (as well, of course, as any
extended-Slater basis set), the natural, and the
hydrogenic functions have been chosen so as to be
precise representations of the hydrogen ground
state and hence of the single 1s orbital that it
represents. The Gaussian function, which is
employed to reduce calculating costs in large
computations, is a poor representation of the
hydrogen 1s wave function, which has the form
of a broad-based, concave cone with a pointed
top, whereas the Gaussian function has a convex
shape with a rounded top. From Fig. 2.3 it can be
seen that the representation of the hydrogen
ground-state atom with two or three atom-
optimized Gaussian orbitals is still poorly shaped

Fig. 2.2. The total electron density of the ground state
of the oxygen atom from a large, near-Hartree–Fock
extended-Slater basis set (exten. Sl.); a double-zeta Slater
basis set (2-ζ); a (52) Gaussian; and a minimum-Slater
basis set (min. Sl.). The basal plane of each plot represents a cutting passing through the oxygen nucleus, with
the electron density being shown perpendicular to this
plane. The peak of each plot (the total electron density
at the nucleus) appears at 2.2×10^3 $e/\text{Å}^3$ for min. Sl.,
1.8×10^3 $e/\text{Å}^3$ for (52), and 2.1×10^3 $e/\text{Å}^3$ for exten. Sl.

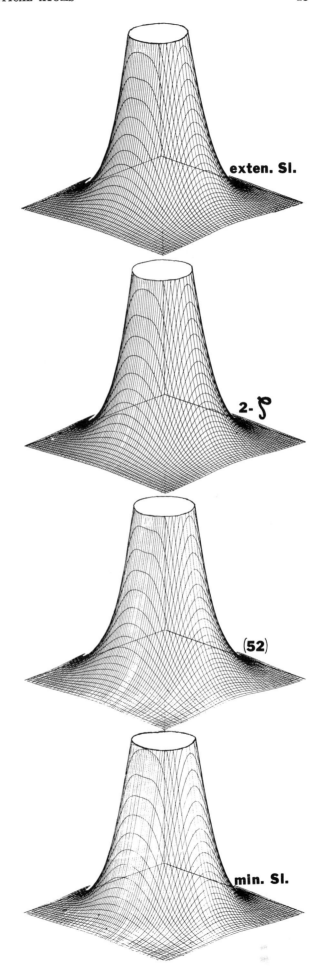

compared to the analytical function (the minimum Slater) and that, although an appropriately pointed top is achieved with atom optimization of 12 s-type Gaussian orbitals, the resulting electron-density plot is still too diffuse to exhibit a sufficiently high electron density in the region of the nucleus. In view of these findings, it should be remembered that in plots of the molecules shown in Chapter 3, the electron densities of the hydrogen atoms are incorrectly represented as having rounded rather than pointed tops.

Typical atomic orbitals of a second-period atom (i.e., $n = 2$) are shown in Fig. 2.4, which presents electron densities calculated for the oxygen atom in its ground state, ^{3}P. In Fig. 2.4A, the electron-density diagrams are based on a double-zeta optimized Slater basis set [1]; in Fig. 2.4B the plots are based on an optimized Gaussian basis set [2] made up of five s-type and two p-type

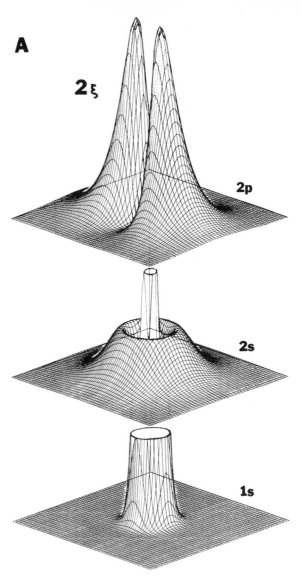

Fig. 2.3. The total electron density (identical to the 1s orbital density) of the ground-state hydrogen atom, as represented in the following atom-optimized basis sets: Sl., a single-Slater function (identical to the precise analytical function); (12), 12 s-type Gaussian functions; (3), three s-type Gaussian functions; and (2), two s-type Gaussian functions.

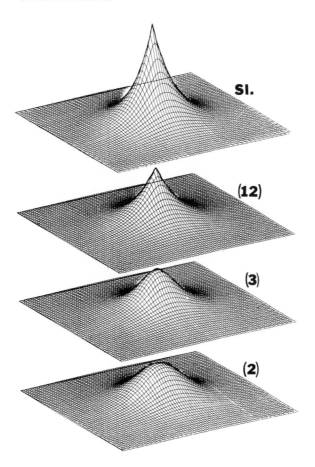

Fig. 2.4. Cross-sectional electron-density plots of the filled atomic orbitals of the ground-state oxygen atom as determined (A) in a double-zeta Slater and (B) in a (52) Gaussian basis set.

functions, i.e., a (52) basis set. The 1s orbital is shown at the bottom of this figure with the 2s coming next and the 2p at the top. The upper portions of the 1s and 2s plots are deleted to save space.

Typical atomic orbitals of a third-period atom (i.e., $n = 3$) are shown in Fig. 2.5, which presents electron densities calculated for the sulfur atom in its ground state, ^{3}P, as obtained from an atom-optimized Gaussian basis set [5] made up of nine s-type and five p-type functions, i.e., a (95) basis set. A plot of the total electron density is shown at the top of this figure, with the individual orbitals arranged beneath it in order of increasing

Fig. 2.4. (continued)

B

(52)

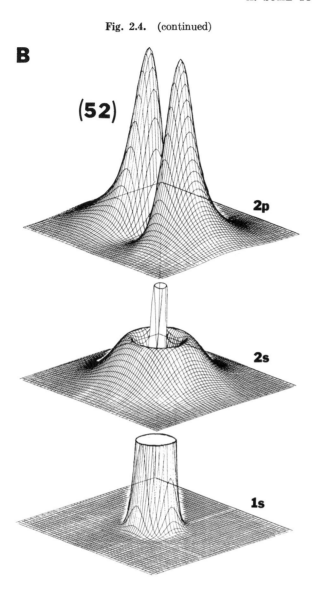

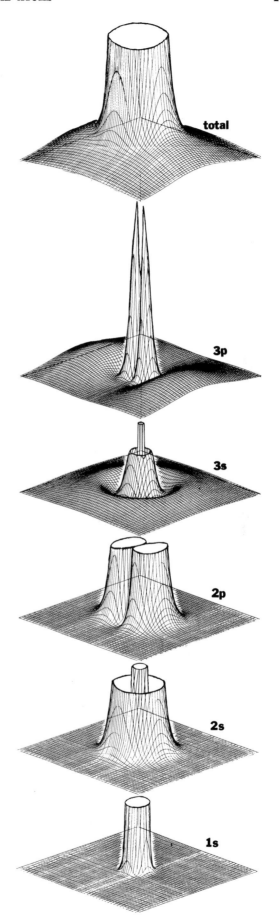

stability from top to bottom. Again, the taller plots are truncated. Note the diffuse character of the outer density maximum (i.e., the outer antinode) of the valence orbitals (the 3s and the 3p). Chemical bonding of the third-period atoms must be effected through overlap of these diffuse outer antinodes, whereas for second-period atoms the bonding involves much more compact valence orbitals, as can be seen in Fig. 2.4. For atoms beyond the third period, the outer antinodes of the valence-shell orbitals are even more diffuse than those shown for sulfur in Fig. 2.5.

Fig. 2.5. Cross-sectional electron-density plots corresponding to the ground-state sulfur atom as determined in a (95) Gaussian basis set. The uppermost plot corresponds to the total electron density and the lower plots to the 3p, 3s, 2p, 2s, and 1s orbitals, respectively.

B. Some Typical Molecules

In the simple localized representation of the water molecule, a representation that is readily characterized by its electron-dot formulation

$$\overset{..}{:}\!\underset{}{\overset{..}{O}}\!:H$$
$$H$$

there are four spin-coupled pairs of electrons, each corresponding to a valence-shell orbital. Two of these localized orbitals represent the unshared pairs of electrons on the oxygen and the remaining two each correspond to an O–H bond. The orbitals obtained from SCF calculations are, however, delocalized. Nevertheless, each of these delocalized orbitals is dominated by a particular electronic function, as can be seen from the electron-density plots of Fig. 2.6. For example, the two most stable of the valence orbitals of water, which are denoted as $2a_1$ and $1b_2$, respectively, are dominated by O–H bonding, with both O–H bonds appearing in each orbital because of delocalization and as required by the symmetry of the molecule. The two filled orbitals of higher energy, $3a_1$ and $1b_1$, are dominated by the oxygen lone-pair character, although orbital $3a_1$ does include some O–H bonding to make up for the oxygen lone-pair character appearing in orbital $1b_2$.

The uppermost electron-density plot in each of the four columns of Fig. 2.6 represents a near-Hartree–Fock calculation [6] with a balanced basis set, so that these plots should form a good representation of the filled-valence orbitals of the ground-state water molecule in the SCF approximation. From these plots, it is clear that the interactions between the constituent atoms to form the molecule is really only a small perturbation of the unreacted atoms and that the symmetry of the overall molecule (or at least the local symmetry for large molecules) determines the orientation of the nodal planes of the atomic orbitals of which the molecular orbitals partake.

Fig. 2.6. Cross-sectional electron-density plots in several basis sets for the valence orbitals of the water molecule, corresponding to the geometrical plane passing through the three atomic nuclei for all orbitals but $1b_1$, for which the basal plane passes through the oxygen and bisects the line connecting the two hydrogen atoms. A full column is devoted to each orbital, with the top line of each column corresponding to the extended-Slater basis set, the next line to a (1062/42) Gaussian, the third to a (73/3) Gaussian, and the fourth to a minimum-Slater basis set. The fifth line represents the results from a CNDO semiempirical calculation and the bottom line those from a one-center calculation with an extended-Slater basis set assigned only to the oxygen atom.

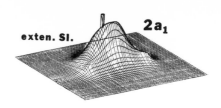

exten. SI. $2a_1$

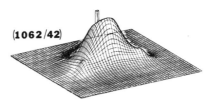

(1062/42)

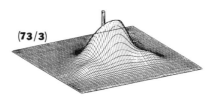

(73/3)

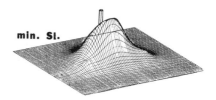

min. SI.

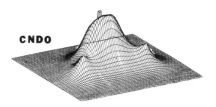

CNDO

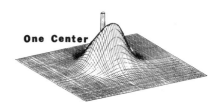

One Center

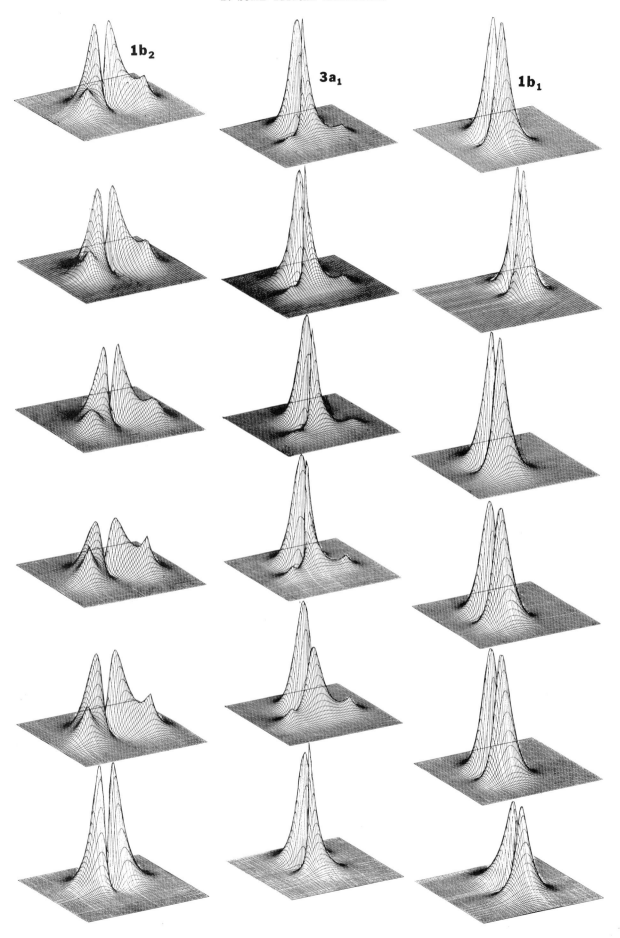

According to the idea (extended *Aufbau* principle) that the more stable atomic orbitals should be utilized first in molecule building, it is not surprising to see that the water molecular orbital of lowest energy, $1a_1$, corresponds to the oxygen 1s core electron and that the most stable of the valence molecular orbitals, $2a_1$, is based on interaction of the oxygen 2s atomic orbital with the 1s orbitals of the pair of hydrogen atoms. The next molecular orbital in order of descending stability, $1b_2$, is based on the oxygen 2p orbital, which is oriented so that each lobe interacts separately with the 1s orbital of a hydrogen. Of the two remaining orthogonal 2p atomic orbitals of the oxygen, one is so oriented that one of its lobes interacts with the pair of hydrogen 1s orbitals; this leads to the molecular orbital, $3a_1$, which also includes oxygen lone-pair character. Finally, the oxygen 2p orbital, which is so oriented that it cannot interact with the hydrogen atoms (since they lie in its nodal plane), makes up the least stable filled orbital, $1b_1$, which corresponds to oxygen lone-pair character in the water molecule.

If we consider the line passing through the oxygen nucleus and halfway between the two hydrogen nuclei as the axis of the molecule, we can discuss the molecular orbitals in terms of their σ and π character with respect to this axis. For example, molecular orbital $2a_1$ may be described as being dominated by O–H $(s_\sigma - s_\sigma)$; $1b_2$ by O–H $(p_\pi - s_\pi)$; $3a_1$ by O–H $(p_\sigma - s_\sigma)$; and $1b_1$ by n_0 $(p_{\pi'})$, where n stands for a lone pair. Note that this use of the σ and π notation is far from standard.

Each of the columns of Fig. 2.6 shows a given valence orbital of the water molecule as represented in six different basis sets. The top plot in each of these columns corresponds to an extended-Slater representation [6] which reproduces bonding orbitals well. In this near-Hartree–Fock

calculation, a single 1s, two 2s, three 2p, and one 3d Slater function have been employed to describe the oxygen, with one 1s, one 2s, and one 2p Slater function for each hydrogen. The next lower course of plots in this figure corresponds to a large Gaussian basis set [7], consisting of ten s-, six p-, and two d-type Gaussian orbitals for the oxygen and four s and two p types for each of the hydrogens, denoted as (1062/42). The plots located third from the top correspond to a smaller Gaussian basis set [8], in which only s and p orbitals have been employed. In this set, seven s- and three p-type Gaussian functions describe the oxygen, with three s types to account for each hydrogen. The fourth row from the top corresponds to the minimum-Slater representation [8] (a single function apiece for the 1s, 2s, and 2p orbitals of the oxygen and for the 1s of the hydrogen), whereas the fifth row shows the results of a CNDO semiempirical calculation [8]. The CNDO calculation deals only with the valence orbitals, using a minimum-Slater basis set and exponents derived from Slater's rules. Furthermore, most of the integrals are neglected and the remainder are estimated semiempirically.

The bottom row of plots in Fig. 2.6 result from a one-center calculation [9] in which two Slater functions were assigned to the 1s, three to the 2s, three to the 2p, two to a 3d, and one to a 4f atomic orbital of the oxygen, with no explicit mathematical description being given for the atomic orbitals of the hydrogens. This one-center representation resulted in a total energy that was considerably better than that from the minimum-Slater calculation and only 0.6 au from the estimated Hartree–Fock limit (see Table 2.I). However, it is apparent from Fig. 2.6 that, although the orbital symmetries have been correctly established in the one-center calculation by the presence of the hydrogen nuclei, there is no congregation of

Table 2.I   Various Calculations of the Total and Valence-Orbital Energies[a] (in au) of Water

Basis set	Total	Orb $2a_1$	Orb $1b_2$	Orb $3a_1$	Orb $1b_1$
Exten. Sl. [6]	−76.005	−1.339	−0.728	−0.595	−0.521
(1062/42) [7]	−76.059	−1.352	−0.719	−0.582	−0.507
(73/3) [8]	−74.623	−1.301	−0.678	−0.532	−0.493
Min. Sl. [8]	−75.703	−1.285	−0.624	−0.466	−0.403
CNDO [8]	—	−1.449	−0.775	−0.696	−0.653
One center [9]	−75.922	−1.326	−0.681	−0.556	−0.495
Experimental [6]	−76.481	—	−0.60	−0.53	−0.46

[a] 1 au of energy = 27.211 eV.

Table 2.II Values for the Dipole Moment of the Water Molecule

Source	Moment
Exten. Sl. [6]	2.04
(1062/42) [7]	2.00
(73/3) [8]	2.32
Min. Sl. [8]	1.92
CNDO [8]	2.08
One center [9]	2.08
Experimental	1.85 ± 0.02

electronic density about the hydrogen nuclei, even though d and f polarizing functions have been employed in the basis set allotted to the oxygen atoms.

From inspection of the electron-density plots of Fig. 2.6, it is apparent that the top two sets of plots (both near-Hartree–Fock) are very similar, except for the expected slight rounding of the tops of the hydrogen atoms in the Gaussian representation of the molecular orbitals. With respect to the electron-density distribution, it appears that the smaller Gaussian representation, (73/3), is the most similar of the remaining descriptions to these near-Hartree–Fock results. Of course, there is a pronounced rounding of the top of the hydrogen atoms in the (73/3) basis set, an effect that shows up particularly in orbital $2a_1$. In contrast,

the minimum-Slater representation seems to lead to a much too pointed electronic distribution in the vicinity of the hydrogen nuclei for orbital $1b_2$. This excessive pointedness is particularly noticeable with the CNDO calculation, where it shows up for orbital $2a_1$ and $3a_1$ as well as for $1b_2$.

In spite of the considerable variation in total energy shown in Table 2.I between these different mathematical representations, all of the orbital-density distributions except that of the one-center calculation are reasonably comparable—with the (73/3) Gaussian basis set giving the best looking results of the limited (sp) representations. Unfortunately, many of the one-electron properties are extremely sensitive to fine details in the electron distribution. This is illustrated in Table 2.II, in which the dipole moments calculated from the same wave functions on which Fig. 2.6 is based are compared with the experimentally measured dipole moment of water. Lest from viewing this table the reader become disappointed in the capability of *ab initio* quantum-chemical calculations to give reasonably good values for experimental properties at the present state of the art, he is referred to Table 2.III, in which experimental values for several other one-electron properties are compared with those calculated from the two wave functions corresponding to the two near-Hartree–Fock calculations shown at the top of Fig. 2.6.

Table 2.III Calculated and Experimental Values of Some One-Electron Properties of Water

Property (units)	Symbol	Extended Slater	(1062/41) Gaussian	Experimental
Quadrupole coupling constants for deuterium (MHz/sec)	$\chi_{\xi\xi}(D)$	0.3626	0.3411	0.3152 ± 0.0077
	$\chi_{\xi\eta}(D)$	−0.0085	−0.0083	−0.0088 ± 0.0087
	$\chi_{\eta\eta}(D)$	−0.1586	−0.1478	−0.1393 ± 0.0070
Angle of rotation for diagonalization of the quadrupole coupling tensor	α	0°57′	0°58′	1°7′ ± 1°10′
Quadrupole coupling constants of ^{17}O (MHz/sec)	$\chi_{\alpha\alpha}(^{17}O)$	−8.33	−8.34	−8.13 ± 0.10
	$\chi_{ab}(^{17}O)$	2.92	2.88	4.33
Component of field gradient perpendicular to the O—D bond	$\eta(O)$	1.48	1.51	0.7 ± 0.1
Hellmann–Feynman forces (au) (electronic contribution)	$f_z(D)$	−1.4975	−1.505	−1.4940
	$f_x(D)$	−2.0482	−2.081	−2.0498
	$f_z(O)$	2.4621	2.928	2.9880
Diamagnetic shielding (ppm)	$\sigma(D)$	102.0	102.9	102.0
Mean square distances of all electrons from the center of mass of $H_2{}^{16}O$ (Å^2)	$\langle \Sigma_k r_{0k}{}^2 \rangle$	5.46	5.37	5.1 ± 0.7
Ionization potentials (au)	I	0.521	0.507	0.463 ± 0.004
		0.595	0.582	0.533 ± 0.011
		0.728	0.719	0.595 ± 0.011

C. Representation of Orbital Shapes in Molecules

Early in the application of quantum mechanics to chemical problems, rough sketches were made to indicate the three-dimensional shapes of the atomic orbitals, particularly those exhibiting nodes. As a familiar example, the following sketch is given to indicate the formation of a π bond from a pair of p atomic orbitals on the atoms of a diatomic molecule.

Such representations of three-dimensional shapes can be extended to electron densities as well as to the wave functions from which they have been calculated. In either case, however, there is the problem of where to draw the encompassing envelope representing the shape, since any wave function and its respective electron density extends in increasingly diminishing amounts out to infinity in all directions.

The problem is well illustrated by the total-electron-density representations of water shown in Fig. 2.7. The top plot of this figure depicts a representation of the electron density in the HOH plane of the water molecule, as calculated from the (73/3) Gaussian basis set [8] used to obtain the orbital electron-density plots forming the third row of Fig. 2.6. By choosing arbitrarily fixed values of the electron density, three-dimensional shape plots of the water molecule are obtained, plots related to cuts parallel to the basal plane of this cross-sectional density plot. These shape plots are shown as the bottom five diagrams of Fig. 2.7. The uppermost shape plot (a) corresponds to a slice through the electron-density distribution plot parallel to its base and intersecting with the electron-density axis at the point marked (a). A similar situation holds for the shape plots labeled (b), (c), (d), and (e). Note that shape plot (a) is just an essentially spherical blob and that plot (e) is also essentially spherical. Thus, the assemblage of three atoms making up the

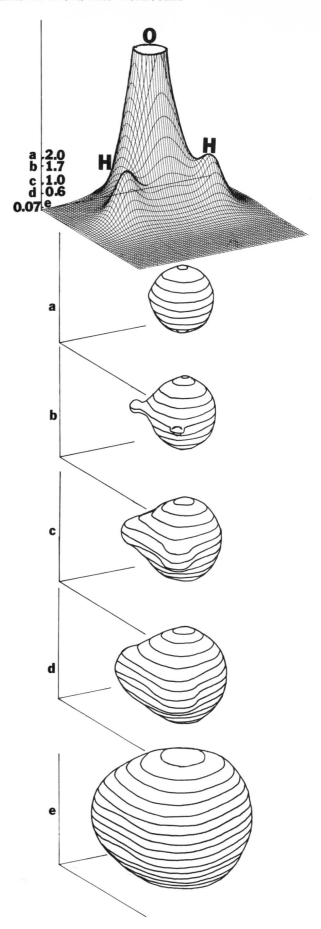

Fig. 2.7. Three-dimensional electron-density shape plots for all the electrons of the water molecule and the relationship of these plots to the cross-sectional electron-density diagram for the total molecule from calculations in a (52/3) Gaussian basis set. This diagram is given at the top of the figure followed by five shape plots. The contour surface in shape plot (a) corresponds to 2.0 $e/Å^3$, in plot (b) to 1.7 $e/Å^3$, in plot (c) to 1.0 $e/Å^3$, in plot (d) to 0.6 $e/Å^3$, and in plot (e) to 0.07 $e/Å^3$.

water molecule shows the best detail at intermediate electron densities.

It is apparent from Fig. 2.7 that the size and detailed shape of a molecule or its constituent orbitals are determined by the value of the electron density used to establish the contour envelope. One way to estimate a physically meaningful value for this electron density is to set it to correspond to the van der Waals radius of the atom or molecule.

In Table 2.IV values of the electron density corresponding to experimental van der Waals radii are presented, as calculated from various wave functions of atoms and molecules—wave functions that have been determined with restricted Gaussian or Slater basis sets. For polyatomic molecules, the standard van der Waals radii given for atoms may be applied to those atoms that make up the outer parts of the molecule. Likewise, because valence orbitals make up the exterior portion of an atom, the usual van der Waals radii may be used to obtain suitable elec-

tron densities for the individual valence orbitals. From the data given in this table, as well as from similar comparisons, we have decided to employ an electron density of 0.6 $e/Å^3$ for delineating the shape plots of the total atoms or molecules described below and 0.14 $e/Å^3$ for the constituent valence orbitals.

Using this value of 0.14 $e/Å^3$, electron-density shape plots have been made for the four valence molecular orbitals of water (see Fig. 2.8). It is interesting to compare these with the orbital shape plots given in *The Organic Chemist's Book of Orbitals*, by Jorgenson and Salem [10], for the water molecule. The major difference between the orbital shape plots calculated by these authors [from extended-Hückel wave functions] and our plots (Fig. 2.8) for the respective electron densities [from *ab initio* (73/3) Gaussian computation] is found in orbital 1b$_2$, the lobes of which appear egglike in the Jorgenson–Salem representation, and in orbital 3a$_1$, the front lobe of which is more of a prolate ellipsoid and the rear lobe of which is

Table 2.IV Calculated Electron Densities Corresponding to the van der Waals Radii

| System | Basis set[a] | Electronic densities ($e/Å^3$) | | | van der Waals radius (Å) |
		Total	Valence "s"	Valence "p"	
H atom	2 GTO	0.115	0.115		1.20
H atom	3 GTO	0.108	0.108		
H in C_2H_6	Min. STO	0.270	0.142		
H in C_2H_4	73/2 GTO	0.148	0.115		
H in C_2H_2	Min. STO	0.202	0.067		
H in H_2O	73/3 GTO	0.202	0.067		
H in H_2S	95/3 GTO	0.142	0.067		
H in BH_3	Min. STO	0.162	0.067		
C atom	Min. STO	0.405	0.162	0.121	1.70
C atom	52 GTO	0.432	0.189	0.115	
C atom	Double-zeta STO	0.385	0.169	0.108	
C in C_2H_6	Min. STO	0.574	0.169	0.148	1.70
C in C_2H_4	73/2 GTO	0.742	0.115	0.155	
C in C_2H_2	Min. STO	0.607	0.270	0.121	
C in C_2H_4S	95/52/3 GTO	0.607	0.142	0.148	
C in CH_3PH_2	95/52/3 GTO	0.945	0.169	0.202	
N atom	52 GTO	0.472	0.189	0.142	1.50
N atom	Double-zeta STO	0.405	0.169	0.121	
O atom	52 GTO	0.574	0.148	0.277	1.40
O atom	Double-zeta STO	0.472	0.142	0.223	
P atom	95 GTO	0.405	0.169	0.115	1.90
P in CH_3PH_2	95/52/3 GTO	0.607	0.115	0.202	1.90
S atom	95 GTO	0.594	0.162	0.283	1.85
S in SH_2	95/3 GTO	0.574	0.169	0.148	1.85
S in C_2H_4S	95/52/3 GTO	0.810	0.101	0.094	

[a] GTO stands for Gaussian-type orbital and STO for Slater-type orbital.

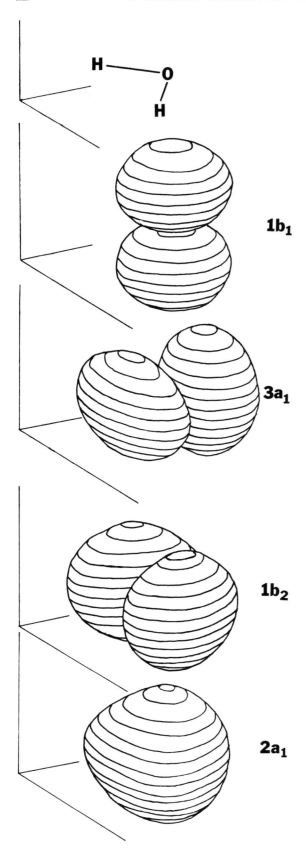

Fig. 2.8. Three-dimensional electron-density shape plots for the various valence orbitals of the water molecule, with the contour surfaces corresponding to 0.14 $e/\text{Å}^3$. Based on a (52/3) Gaussian representation.

more spherical in their representation than in our pictographs.

Another simple molecule for which shape plots are easily visualized is acetylene. However, before the shape plots are shown, it is desirable to present the cross-sectional electron-density distributions for this molecule. These are given for a minimum-Slater calculation [11] in Fig. 2.9 for the total electrons (at the top) and then the valence-shell electrons, followed by the five filled valence-shell orbitals ordered with respect to energy, with the most stable being at the bottom. Note that the chief difference between the shape of the cross-sectional total density and the cross-sectional valence-shell density is the appearance of nodal rings around the positions of the two carbon nuclei of the latter. These annular diminutions in electron density are attributable to the necessity that the valence shell be orthogonal to the core shell of each carbon. Again, the standard pattern of ordering of the orbital energies, as described on page 18 for the water molecule, is seen to emerge for acetylene, for which the most stable valence-shell molecular orbital, $2\sigma_g$, is seen to involve the s atomic orbitals of all four atoms; with the next orbital, $2\sigma_u$, also based on the s atomic orbitals, being antibonding for the pair of carbon atoms and bonding for the two C–H bonds. Following this, orbital $3\sigma_g$ involves the carbon p orbitals interacting with each other and with the terminal hydrogen atoms to cause σ bonding for the full length of the molecule. Finally, there is a degenerate pair of π-type molecular orbitals, $1\pi_u$, which represent the least stable electrons in the ground-state acetylene molecule. These are typical, equally filled π orbitals corresponding to sidewise interaction between each pair of 2p atomic orbitals (per carbon atom) that lies perpendicular to the axis of this linear molecule. One member of this degenerate pair of $1\pi_u$ orbitals appears as shown; the second looks the same for a plane passing through the HCCH molecule at right angles to the plane depicted. In the depicted plane, this latter orbital can, of course, show no electron density, because this is its nodal plane.

The corresponding shape plots for the acetylene molecule are shown in Fig. 2.10. In the shape-plot representation of the total electron density, the two electrons from the pair of hydrogen atoms are seen to be essentially buried in the 12 electrons coming from the two carbon atoms. For the average electron density chosen to correspond to the van der Waals radii, it is seen that molecular orbital $2\sigma_g$ shows up as a sphere with dimples on opposite sides and with two tiny balls of charge within these dimples surrounding the hydrogen nuclei. Orbital $2\sigma_u$ in this representation consists of two large balls surrounding the C—H bonding regions of the molecule with a dimple in each ball at the nearest carbon atom, which shows up as a small ball within the dimple. Molecular orbital

$3\sigma_g$ shows up as three balls in line and each of the $1\pi_u$ or $1\pi'_u$ orbitals really looks like a pair of thick sausages, according to the "sausage-orbital" sobriquet which has long been used to describe π bonding.

Fig. 2.9. Cross-sectional electron-density plots for the acetylene molecule (calculated in a minimum-Slater basis set), with the basal plane passing through the nuclei of the constituent atoms. The lower four plots represent the molecular orbitals, with the bottom one ($2\sigma_g$) being the most stable.

Fig. 2.10. Electron-density shape plots for the acetylene molecule. The top diagram corresponding to the total molecule is shown at 0.6 $e/\text{Å}^3$ and the remaining orbital-density plots at 0.14 $e/\text{Å}^3$. These shape plots correspond to the cross-sectional electron-density plots of Fig. 2.9.

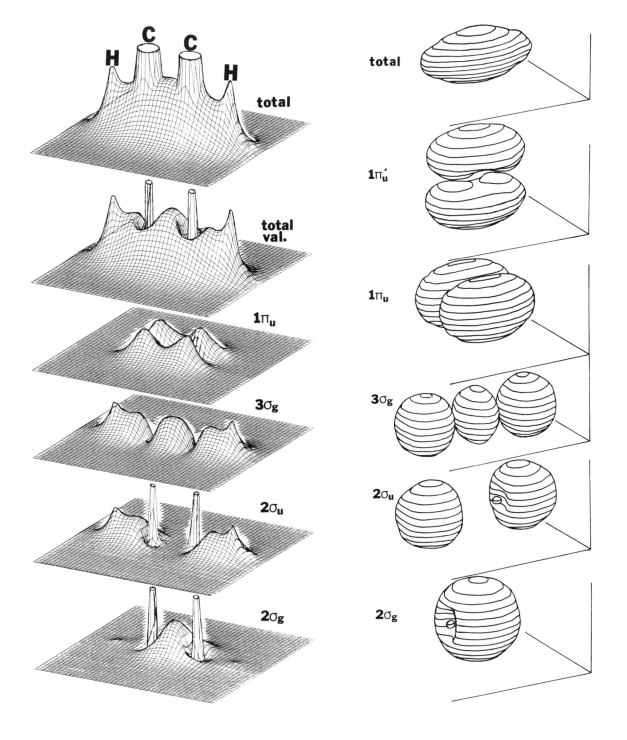

D. Effect of Poor Basis Sets on Electron-Density Distributions

Although many inappropriate basis sets have been employed in calculations appearing in the literature, their choice has been made through error. Moreover, it is usually difficult to compare these wave functions with reasonably equivalent functions that do not suffer from their deficiencies. Therefore, a few years ago we published a study [12] in which the difluoromethane molecule was investigated in a number of basis sets, some of which were badly unbalanced while others represented "starved" representations in which an in-

sufficient number of Gaussian functions were employed.

To show the effect of unbalanced basis sets on the molecule, it is not necessary to exhibit all of the ten valence-shell molecular orbitals. Therefore, Figs. 2.11–2.13 have been limited to molecular orbitals $3a_1$, $5a_1$, and $3b_1$ in the FCF plane and Figs. 2.14 and 2.15 to orbitals $4a_1$ and $2b_2$ in the HCH plane of the CH_2F_2 molecule. As can be seen from Figs. 2.11–2.13, molecular orbital $3a_1$ corresponds to F–C $(s_\sigma–s_\sigma)$ bonding, with orbital $5a_1$ approximating F–C $(p_\pi–p_\pi)$ bonding and $3b_1$ exhibiting F–C $(p_\sigma–p_\sigma)$ bonding. Likewise, in Figs. 2.14 and 2.15, molecular orbital $4a_1$ is seen to be

Fig. 2.11. Cross-sectional electron-density plots of molecular orbital $3a_1$ in the CF_2 plane of methylene fluoride. The top plot represents an acceptable Gaussian basis set, (73/73/3), whereas the middle two plots are based on highly unbalanced basis sets, (52/73/3) and (73/52/3). The bottom plot corresponds to a Gaussian basis set, (31/31/1), that is much too small to give an acceptable representation of the molecule.

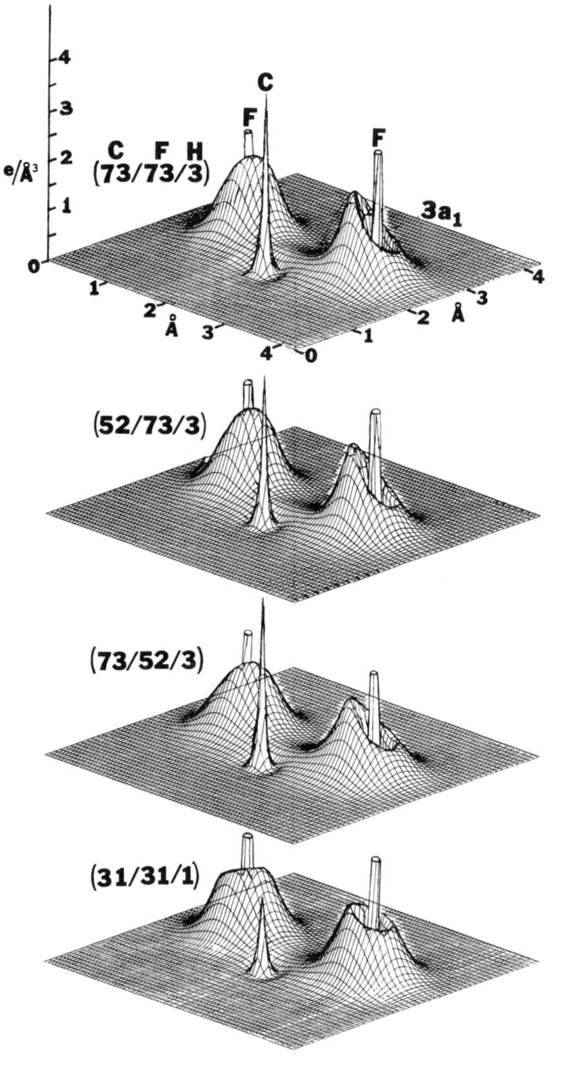

Fig. 2.12. Cross-sectional electron-density plots of molecular orbital $5a_1$ in the CF_2 plane of methylene fluoride, using the same basis sets as in Fig. 2.11.

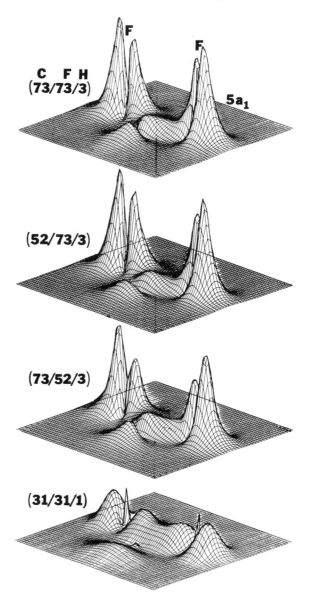

dominated by C–H (s_σ–s_σ) bonding and $2b_2$ to C–H (p_σ–s_σ) bonding. The layout of Figs. 2.11–2.15 is similar to that of Fig. 2.6 in that each column of plots corresponds to a different molecular orbital, with the various rows of the plots corresponding to different basis sets. Note in the bottom row that the starved basis set in which three Gaussian s-type and only one p-type function is allotted to each

second-period atom (C or F), with one s-type Gaussian to each hydrogen, gives a very poor representation of all of the molecular orbitals except for orbital $3a_1$ (see Fig. 2.11). At least superficially, the other basis sets seem to lead to reasonably well-shaped electron-density plots.

The top diagram in Figs. 2.11–2.15 corresponds to a (73/73/3) Gaussian basis set, which gives a reasonably satisfactory description of the molecule. Indeed the total energy for difluoromethane obtained from this calculation was −237.51 au, which should be compared with the value of

Fig. 2.13. Cross-sectional electron-density plots of molecular orbital $3b_1$ in the CF_2 plane of methylene fluoride, using the same basis sets as in Fig. 2.11.

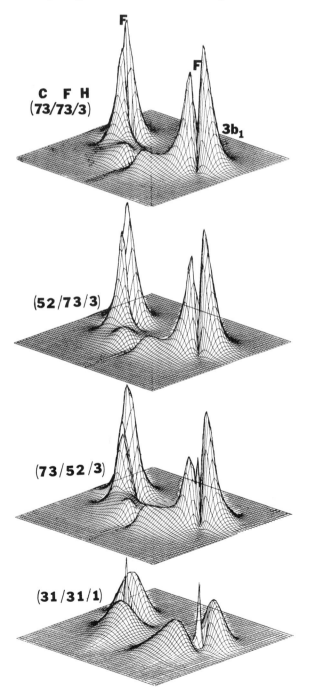

Fig. 2.14. Cross-sectional electron-density plots of molecular orbital $4a_1$ in the CH_2 plane of methylene fluoride, using the same basis sets as in Fig. 2.11.

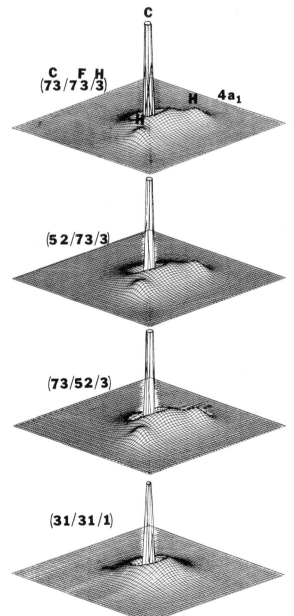

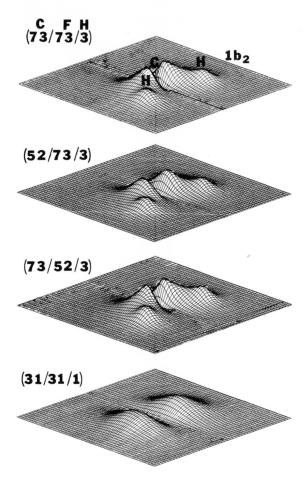

C F H
(73/73/3)
1b₂

(52/73/3)

(73/52/3)

(31/31/1)

Fig. 2.15. Cross-sectional electron-density plots of molecular orbital 1b₂ in the CH₂ plane of methylene fluoride, using the same basis sets as in Fig. 2.11.

−237.87 au obtained from a recent near-Hartree–Fock (sp) computation [13], in which several Gaussian orbitals were employed to emulate the pair of Slater orbitals in a "double-zeta" (sp) basis set.

The second layer of diagrams in Figs. 2.11–2.15 corresponds to an unbalanced basis set describable as (52/73/3) in the order C/F/H. This basis set is probably not an extremely poor representation from the viewpoint of the overall electron distribution, for it allows a larger set of basis functions to the highly electronegative fluorine than to the carbon. The set of diagrams third from the top of Figs. 2.11–2.15 corresponds to a (73/52/3) basis set, which is, of course, badly out of balance. As previously noted, the bottom diagram of each of these figures shows the results obtained using a (31/31/1) basis set, which not only gives a very poor representation for the carbon and fluorine atoms but also leads to abnormal broadening of the hydrogen density distribution. Note that for this starved basis set that the hydrogen does not show up at all, as it should in the HCH plane of Figs. 2.14 and 2.15.

The effect of changing the basis set on difluoromethane [12] is given in Table 2.V for the total energy, the atomic charges (as estimated from a Mulliken population analysis), and the dipole moment. In this table, results are also presented from semiempirical INDO, CNDO, and extended-Hückel (without reiteration) calculations on the difluoromethane molecule.

Table 2.V Effect of Changing the Basis Set on Some Properties of the Difluoromethane Molecule [12, 13]

Basis set C/F/H	Total SCF energy (au)	Atomic charges[a] C	F	H	Dipole moment (debyes)
(31/31/1)	−222.95	−0.67	+0.30	+0.03	−2.24
(31/31/2)	−223.21	−0.35	+0.29	−0.11	−2.40
(52/52/2)	−234.79	+0.06	−0.16	+0.13	1.55
(52/52/3)	−234.96	0.00	−0.17	+0.17	1.85
(73/52/3)	−235.05	−0.04	−0.15	+0.17	1.69
(52/73/3)	−237.16	+0.16	−0.33	+0.25	2.91
(73/73/3)	−237.51	+0.28	−0.40	+0.26	2.72
((105/105/4))[b]	−237.87	+0.33	−0.34	+0.18	2.81
INDO	—	+0.57	−0.24	−0.04	1.96
CNDO	—	+0.40	−0.19	−0.01	1.94
EXTHUC[c]	—	+1.09	−0.71	+0.16	—
Exper.					1.96

[a] These calculations of the formal property called the "charge" on the atom are based on the Mulliken population analysis.

[b] A (105/105/4) Gaussian basis set contracted to a (42/42/2) set, so as to simulate a double-zeta Slater basis set [13].

[c] Extended Hückel without reiterations.

REFERENCES

1. E. Clementi and D. L. Raimondi, *J. Chem. Phys.* **38,** 2686 (1963).
2. D. R. Whitman and C. J. Hornback, *J. Chem. Phys.* **51,** 398 (1969).
3. E. Clementi, *J. Chem. Phys.* **40,** 1944 (1964).
4. E. Clementi, C. C. J. Roothaan, and M. Yoshimine, *Phys. Rev.* **127,** 1618 (1962); also see E. Clementi, "Tables of Atomic Functions" [supplement to a paper in *IBM J. Res. Develop.* **9,** 2 (1965)]. IBM Res. Lab., San Jose, California, 1965.
5. J. R. Van Wazer and I. Absar, *Advan. Chem. Ser.* **110,** 20 (1972).
6. S. Aung, R. M. Pitzer, and S. I. Chan, *J. Chem. Phys.* **49,** 2071 (1968).
7. D. Neumann and J. W. Moskowitz, *J. Chem. Phys.* **49,** 2056 (1968).
8. I. Absar and J. R. Van Wazer, unpublished calculation.
9. R. Moccia, *J. Chem. Phys.* **37,** 910 (1962).
10. W. L. Jorgensen and L. Salem, "The Organic Chemist's Book of Orbitals." Academic Press, New York, 1973.
11. W. E. Palke and W. N. Lipscomb, *J. Amer. Chem. Soc.* **88,** 2384 (1966).
12. M. L. Unland, J. H. Letcher, I. Absar, and J. R. Van Wazer, *J. Chem. Soc., A* p. 1328 (1971).
13. C. R. Brundle, M. B. Robin, and H. Basch, *J. Chem. Phys.* **53,** 2196 (1970); also see L. C. Snyder and H. Basch, "Molecular Wave Functions and Properties," pp. T-274 to T-279. Wiley (Interscience), New York, 1972.

3 Cross-Sectional Plots of Electron Densities

A. Introduction

The purpose of this chapter is to acquaint the reader with the appearance of cross-sectional electron-density plots of the filled molecular orbitals for a variety of molecules. In the next three sections (B–D) of this chapter, the molecular orbitals of some small organic molecules are presented to give the reader a feeling for the subject. This is followed (Section E) by a discussion of the molecular orbitals in two extremely polar small molecules. The following three sections (F–H) deal with the comparison of related molecular orbitals between molecules which differ only in having one of the constituent atoms substituted by the next larger atom from the same group of the periodic table. Then, in Section I, the molecular orbitals of small cyclic molecules are exemplified by cyclopropane and its two analogs in which one of the methylene groups is substituted by a P—H or an S bridge.

Correlation of related molecular orbitals is continued in Section J with a series of molecules in which the first and then the second unshared pair of electrons on hydrogen sulfide is coordinated to an oxygen atom. This same procedure of substituting an unshared pair of electrons by a bond to an oxygen atom is looked at further in Section K for some phosphorus compounds, with the added twist of also substituting a fluorine for a hydrogen. The demonstration of the interrelationship between similar orbitals in disparate molecules is completed with Section L, in which the molecular orbitals of water, formaldehyde, and ketene are contrasted and shown to have much in common.

In two following sections (M and O), the emphasis shifts to orbital electron densities corre-

sponding to the molecular-orbital description of a molecule exhibiting three-center bonds or a molecule having a coordination number larger than four, a value that represents the highest number of bonds which could be accounted for by the use of only s and p orbitals in classical valence theory. This is joined in Section N by a discussion of the effect of the addition of d atomic orbitals to the LCAO description of third-period atoms, with emphasis on d-orbital participation in back bonding. Finally, in Section P the effect of internal rotation on electron densities is demonstrated.

B. Methane, Ethane, and Ethylene

CH_4—Because of its tetrahedral symmetry, the methane molecule is difficult to envisage in terms of its four filled valence-shell molecular orbitals: the $2a_1$ orbital and the degenerate set of three $1t_2$ orbitals. In Fig. 3.1 the electron-density plots (derived from a minimum-Slater calculation [1]) are presented for a cross-sectional cut passing through the nuclei of the central carbon and two of the hydrogen atoms. Remember that six necessarily equivalent cuts of this type can be made through a single methane molecule. The top two plots of Fig. 3.1 show the total electron density and the electron density of the valence shell. As is always the case with atoms of the second period of the periodic table, the valence shell must exhibit a node to keep it orthogonal with the core, which is molecular orbital $1a_2$ of methane (i.e., the carbon 1s orbital). This results in the annular concavity which appears in the valence-shell electron-density plot encircling the carbon nucleus.

Molecular orbital $2a_1$, being the most stable of the filled valence-shell orbitals, is shown as the bottom diagram of the figure. It exhibits no nodal planes and must therefore correspond to the overlap of the 2s atomic orbital of carbon with the 1s orbitals of each of the four hydrogen atoms. If the electron density is plotted on a plane run through the carbon nucleus and bisecting the line passing through the two hydrogen atoms of the plot of Fig. 3.1, it would pass through the two hydrogen atoms which lie above and below the plane of orbital $2a_1$ as represented in the figure. This perpendicular plane for orbital $2a_1$ would then look exactly like the diagram shown in the figure except for its changed orientation in space.

Electron densities corresponding to a set of three $1t_2$ orbitals, each of which must exhibit exactly the same energy (i.e., they are degenerate) are shown as the upper three valence-orbital diagrams in Fig. 3.1. These diagrams indicate that each of these degenerate orbitals exhibits a nodal plane and that these planes are mutually perpendicular. This is expected from their formation from each of the three 2p orbitals of the carbon interacting with the appropriate hydrogen s orbitals. The lower $1t_2$ diagram corresponds to the situation in which the 2p lobe of the carbon intersects the pair of hydrogen atoms in the chosen plane, thereby making a kind of three-center bond, whereas the upper two $1t_2$ orbitals correspond to a pair of π-like interactions, with reference to the line passing through the carbon atom and the midpoint between the hydrogens. Note that in the perpendicular plane passing through the carbon atom and bisecting the hydrogen pair, the uppermost of the set of three degenerate $1t_2$ orbitals is found to appear identical to the one pictured below it but now involving the two other hydrogen atoms of the molecule.

A three-dimensional sketch of the four valence-shell molecular orbitals calculated for methane is shown in Fig. 3.2 in order to clarify their spatial

Fig. 3.1. Cross-sectional electron-density plots of methane, showing a plane passing through the carbon and two of the hydrogen atoms. The basal plane of each graph contains these three atoms; the vertical elevation at any point is proportional to the electron density at that point.

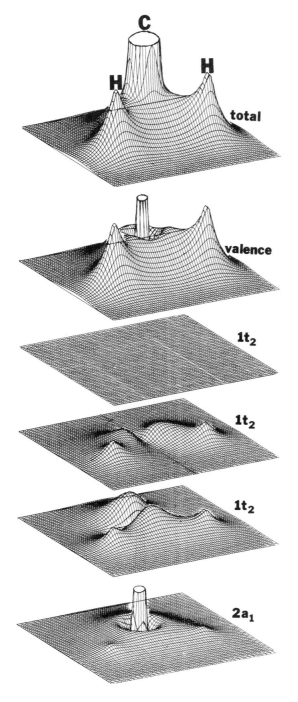

Fig. 3.2. Rough sketch of the shapes of the valence orbitals of methane. The orbitals are numbered to correspond to the graphs of Fig. 3.1, starting at the top of this figure. Numbers 3–5 correspond to the $1t_2$ orbitals and number 6 to orbital $2a_1$. Note that the nodal planes are shown in sketches 3–5 as extending beyond the cube drawn so as to have its alternate corners passing through the four hydrogen nuclei.

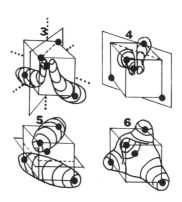

arrangement. Note in Fig. 3.2 that the nodal planes of the three $1t_2$ orbitals are mutually perpendicular. Furthermore, these three degenerate orbitals are constructed from the $2p_x$, $2p_y$, and $2p_z$ orbitals of the carbon atom. A set of three p-type atomic orbitals exhibits nodal planes that are mutually perpendicular; but, since an atom is spherically symmetrical, it is immaterial how the Cartesian coordinate axes (which determine the direction of the nodal plane of each of the p atomic orbitals) are oriented for the free atom. However, when the carbon atom is tetrahedrally surrounded by the four hydrogen atoms in methane, the orientation of the hydrogen atoms with respect to the coordinate axes does make a difference, because this determines the orientation of the nodal planes of the carbon 2p atomic orbitals and therefore the way each of them can interact with the hydrogens. Although the angular position of the molecule, with respect to the Cartesian coordinate system, will determine the shape of the individual orbitals of a degenerate set, the shape of the sum of these orbitals will be independent of the axial orientation. It may be shown, moreover, that a set of degenerate orbitals corresponding to any given orientation may be mathematically derived from a set for any other orientation.

In general, a set of degenerate orbitals is invariant to rotation, although the individual orbitals are not. The shape of the three $1t_2$ orbitals given in Fig. 3.2 corresponds to the molecular orientation employed in the SCF calculation on which Fig. 3.1 is based. Obviously, the sum of the three 2p orbitals of the carbon must interact equally with each hydrogen atom so that the combination of the $2a_1$ and the three $1t_2$ orbitals is equivalent to the four equivalent, tetrahedrally directed $s^{\frac{1}{4}}p^{\frac{3}{4}}$ orbitals used when the methane molecule is discussed in terms of sp^3 hybridization of the carbon atom.

C_2H_6—Again using a minimum Slater basis set [1], the ethane molecule, H_3CCH_3, has been studied in its most stable conformation, the staggered form. As shown in Fig. 3.3, the basal plane of the electron-density plots passes through the two carbon atoms and a pair of opposing (trans) hydrogens, one on each carbon.

The most stable of the valence-shell molecular orbitals, $2a_{1g}$, is seen to have no nodes except for the spherical one that surrounds each carbon atom and acts mutually to orthogonalize the carbon core and the valence-shell orbitals. Thus, this molecular orbital corresponds to the overlap of the valence s orbitals on all of the atoms in the molecule. The next most stable orbital $2a_{2u}$, has

a single nodal plane in addition to these two spherical nodes; it again involves all of the s orbitals in the molecule but with antibonding in the C–C bond, as shown by the region of zero

Fig. 3.3. Cross-sectional electron-density plots of the valence-shell orbitals of the ethane molecule in the *staggered* configuration. The basal plane of the plot passes through the two carbon atoms and a pair of *trans* hydrogen atoms.

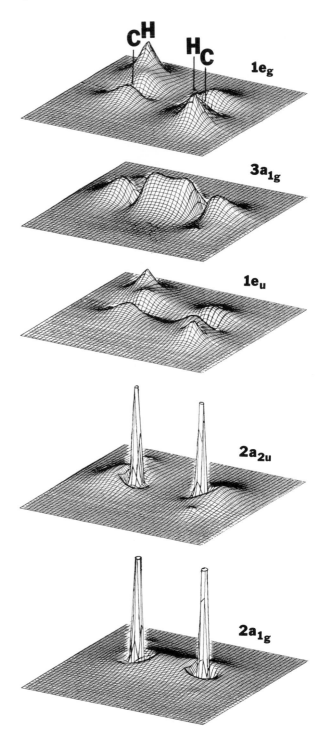

electron density corresponding to the nodal plane that bisects this bond. This orbital is followed by the degenerate pair of $1e_u$ orbitals, which together add triple-bond character to the C–C bond by the pair of mutually perpendicular nodal planes passing through this bond. The pair of $1e_u$ molecular orbitals corresponds to the sideways overlap of the carbon $2p$ orbitals, for which the lobes extend perpendicular to the molecular axis. In this pair of orbitals, there is considerable overlap between these $2p$ lobes and the hydrogen atoms, as indicated by the relatively high electron density in the C–H bonding regions of the density diagram shown in Fig. 3.3. The pair of $1e_u$ orbitals may therefore be considered to exhibit $(p_\pi - p_\pi)$ bonding with respect to the C–C bond and $(p_\sigma - s_\sigma)$ bonding with respect to each of the C–H bonds.

The next orbital in order of decreasing stability is $3a_{1g}$, which exhibits two parallel nodal planes and corresponds to the utilization of the carbon $2p$ orbitals, exhibiting lobes that lie on the C–C bond axis. This results in a four-center-like bond between the carbon and the three hydrogen atoms, as well as in σ bonding between the pair of carbon atoms. The outermost valence orbitals correspond to the degenerate pair labeled $1e_g$, which together add triple-antibonding character to the C–C bond. Each of these orbitals exhibits a mutually perpendicular pair of nodal planes. For each orbital, the nodal plane containing the C–C bond axis leads to C–C π character, whereas the nodal plane bisecting this axis confers antibonding character to the C–C bond, which may therefore be described as $(p_\pi - p_\pi)^*$. These $1e_g$ orbitals also correspond to C–H bonding, with this bonding again involving the pair of carbon $2p$ orbitals for which the lobes extend perpendicular to the C–C bond axis. This C–H overlap involves the interaction of the ring of charge around the carbon atom with the charge on each hydrogen, giving C–H $(p_\sigma - s_\sigma)$ bonding.

It has been common practice to show electron-density difference plots for the formation of a molecule from its ground-state atoms. This type of plot is shown for the ethane molecule in the top two diagrams of Fig. 3.4. The uppermost of these diagrams is a "transparent" plot in which the front and back as well as the upper and lower surfaces of the mesh are depicted; whereas the second plot is the more usual kind, giving a conventional view of the top surface. These two plots correspond to the same cross-sectional plane used for ethane in Fig. 3.3 but the plotted electron density corresponds to that of the C_2H_6 molecule minus the electron density obtained by placing

Fig. 3.4. Transparent and regular cross-sectional electron-density difference plots for all of the electrons of the ethane molecule, corresponding to the same cross-sectional plane used for this molecule in Fig. 3.3. Plot (a) corresponds to the molecular electron density minus that obtained by summing the ground-state constituent atoms placed in the same position in which they appear in the molecule. Plot (b) is a similar one, in which the carbon atoms have been hybridized to the sp^3 configuration. The top plot under either (a) or (b) is transparent and projects below as well as above the chosen plane.

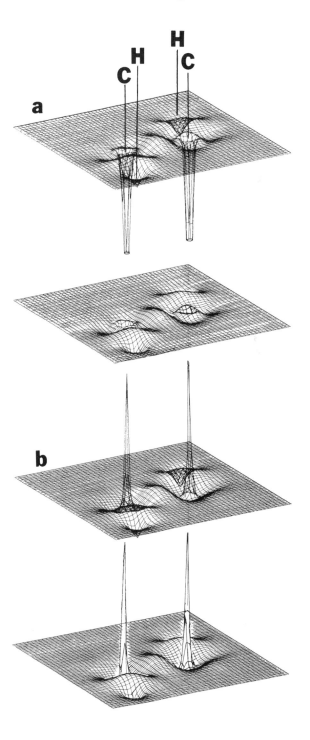

the two carbons and the six hydrogens (as spherically symmetrical ground-state atoms) in the same positions in which these atoms appear in the molecule and adding up the total electron density at each point on the plane due to the interpenetrating neutral atoms. Since the molecular calculation was done with an atom-optimized minimum-Slater basis set, the individual atoms were described by exactly the same atom-optimized Slater basis functions. The lower two diagrams of Fig. 3.4 also represent the difference in electron density between the ethane molecule and its atoms, with the transparent plot again being on top of the regular one. In this case, however, the carbon atoms have been hybridized to the sp³ configuration, employing the same Slater exponents used for the molecule.

In both of the difference plots in Fig. 3.4, the flat humps between each hydrogen and its neighboring carbon and between the two carbon atoms show the amount of charge drawn from the atoms into the bonding region of the molecule. Note that these humps in the bonding regions are about the same whether the carbon atoms are present in their ground state or as the sp³ hybrid. However, the most prominent features in these plots are the sharp upward or downward pointing peaks centered at the positions of the atomic nuclei. The negative cones at the hydrogens are attributable to the fact that the hydrogen atoms in all of these plots exhibit a single electron (i.e., s$^{1.0}$), whereas in the molecule, because of charge transfer, there is only 0.88 e associated with each hydrogen (s$^{0.88}$). Likewise, the valence shell of the ground-state carbon atom is s$^{2.00}$p$^{2.00}$ and in the sp³ hybrid it is s$^{1.00}$p$^{3.00}$ as compared to the molecule, which is s$^{1.24}$p$^{3.13}$, as calculated in the minimum-Slater basis set. This means that the difference plot for the molecule minus atom (ground state) will show a deficiency on each atom of $2.00 - 1.24 = 0.76$ 2s electrons, which leads to the large negative peaks on the carbon. By converting the carbon atoms to the sp³ hybrid, there has been an overcompensation of the carbon 2s orbitals by the amount of $1.24 - 1.00 = 0.24$ e and this causes the peaks at the carbon nuclei in the two lower diagrams of Fig. 3.4 to be positive but considerably smaller than the corresponding negative peaks in the upper two diagrams of this figure. Similar arguments may be made for the carbon 2p orbitals involved in the molecular plane for which the cross-sectional cuts of Fig. 3.4 apply.

C₂H₄—The electron-density plots for the ethylene molecule, H₂CCH₂, which was also calculated

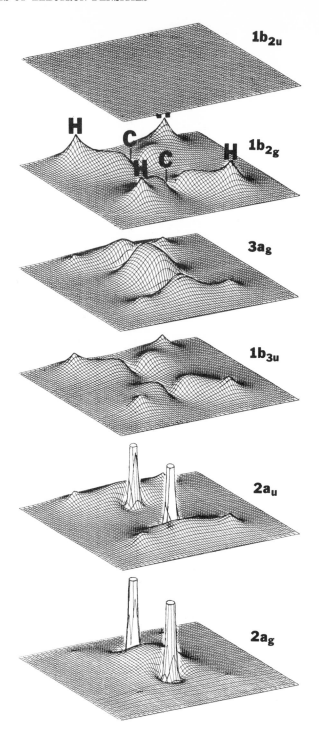

Fig. 3.5. Cross-sectional electron-density plots of the valence-shell orbitals of ethylene in which the basal plane passes through the nuclei of all of the atoms.

in a minimum-Slater basis set [1], are shown in Fig. 3.5, in which the basal plane of each plot passes through the nuclei of all six atoms of this molecule. The most stable valence-shell molecular

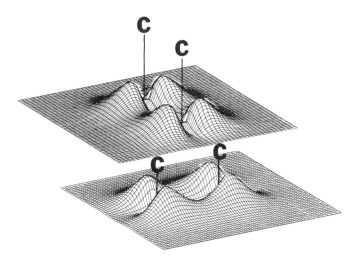

Fig. 3.6. Two views at different angles of the electron distribution of the π-bonding orbital of ethylene, orbital 1b$_{2u}$. The basal planes of the diagrams in this figure are perpendicular to the plane of the molecule shown in Fig. 3.5.

orbital, 2a$_g$, of ethylene, as usual, has no nodal surfaces in the valence region and corresponds to the overlap of the valence s orbitals of all of the atoms in the molecule. The next higher orbital, 2a$_u$, exhibits the nodal plane corresponding to antibonding between the pair of carbon atoms, coupled with considerable C–H bonding involving s atomic orbitals. This is followed by orbital 1b$_{3u}$, in which the carbon 2p lobes in the molecular plane are seen to lead to σ bonding between the carbon and each neighboring hydrogen while, at the same time, contributing π character to the C–C bond. Orbital 3a$_g$, which is the next in order of decreasing stability, corresponds to the use of the carbon 2p orbitals exhibiting nodes lying on the C–C bond axis. As can be clearly seen, this orbital contributes to the C–C σ bond as well as to the C–H bonding. Orbital 1b$_{2g}$ is not unlike orbital 1b$_{3u}$, except that the former is antibonding and therefore has two perpendicular nodal planes, as compared to the single one of the latter, which passes through the C–C bond and is perpendicular to the molecular plane (cf. orbitals 1e$_g$ and 1e$_u$ of ethane in Fig. 3.3). The least stable of the filled molecular orbitals in the ethylene molecule corresponds to the C–C π bond, with no gathering of electron density in the region of the hydrogen atoms. This means that the molecular plane of orbital 1b$_{2u}$ must be a nodal plane corresponding to zero electron density throughout. In order to familiarize the reader thoroughly with the shape

of π-type molecular orbitals, two views of this orbital, 1b$_{2u}$, are shown in Fig. 3.6. In this figure the basal planes of the plots are at right angles to the planes of those in Fig. 3.5, with the basal planes of Fig. 3.6 intersecting those of Fig. 3.5 along the C–C bond axis.

Interrelationships

The molecular orbitals of methane, ethane, and ethylene have been discussed above primarily in terms of the atomic orbitals from which they are constructed according to the LCAO approximation. However, we may consider the molecular orbitals as entities themselves. From this point of view, the gross geometry of the orbitals is determined by the placement of the various nuclei involved in the molecule. Therefore, the orbitals

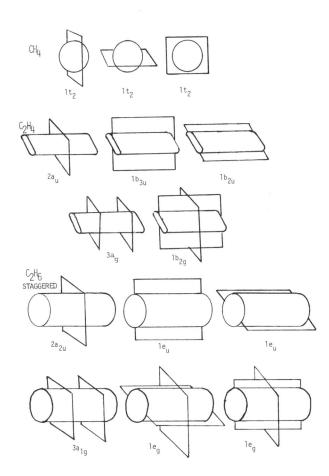

Fig. 3.7. Diagrammatic representations of the nodal surfaces extending into the valence-shell region for the valence orbitals of methane, ethylene, and ethane. The methane molecule is indicated as a ball, the ethylene as a bar, and the ethane as a cylinder for the purposes of this plot.

of the methane molecule must exhibit T_d symmetry, whereas those of ethane and ethylene must have D_{3d} and D_{2h} symmetries, respectively. As is the case with atomic orbitals, the molecular orbitals will exhibit nodal surfaces. The nodal surfaces in the valence shell (i.e., exclusive of the spherical 2s nodes making the 1s and 2s orbitals orthogonal for each carbon atom) of the valence-shell molecular orbitals of CH_4, C_2H_6, and C_2H_4 are shown in Fig. 3.7. Note that the valence-shell orbitals of the highly symmetrical methane exhibit nodal planes not unlike those of the 2s and three 2p orbitals of an atom.

C. Methyl Fluoride, Methyl Chloride, Methyl Isocyanide, and Acetonitrile

The *ab initio* calculations [2] on which the electron-density plots for these molecules are based employed three s-type Gaussian functions for each hydrogen, five s- and three p-types for either carbon, nitrogen, or fluorine, and nine s-, five p-, and one d-type function for the chlorine atom, with all of the s- and p-type Gaussian functions being atom optimized and the d function being set close to its molecularly optimized value. In every case the molecules were laid out so that the basal plane of the cross-sectional electron-density plots passed through the C–X axis of the molecule (where X = F, Cl, NC, or CN) and through one of the methyl hydrogen atoms, which was placed near the upper lefthand corner of each diagram.

In Fig. 3.8 electron-density plots are given for the total electrons in methyl fluoride and methyl chloride. Note the difference in these diagrams between the second-period atom, fluorine, for which the valence electrons are held closely to the core, and the third-period atom, chlorine, for which the valence electrons are relatively diffuse. Since fluorine is considerably more electron withdrawing than carbon, the slope of the electron-density surface measured along the bond axis at a point in the center of the C–F bond is seen to be large and pointing upward toward the fluorine atom. For methyl chloride, the center of the bond axis is flat, even though the Pauling electronegativity of chlorine (3.15) is appreciably larger than that of carbon (2.60). According to the Mulliken gross populations, the charge on the carbon atom of methyl fluoride is $-0.22\ e$, while on the fluorine atom, it is $-0.29\ e$, with these charges being taken from the three hydrogen atoms, whereas for methyl chloride, the charge on the carbon is $-0.45\ e$, and on the chlorine it is $-0.06\ e$. It is interesting to note that in both of these molecules

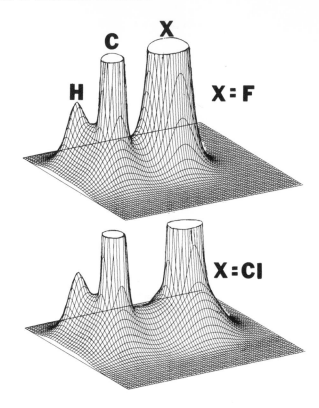

Fig. 3.8. Cross-sectional electron-density plots for all the electrons in methyl fluoride and methyl chloride. The basal planes of these diagrams pass through the carbon, the halogen, and one of the hydrogen atoms.

each of the hydrogen atoms bears a charge of $+0.17\ e$, so that the →CF moiety has (at least in these descriptions) the same electron-withdrawing power as the →CCl moiety. This conclusion from the population analysis is borne out by the exact duplication of the shape of the total electron-density distribution around the hydrogen nucleus in Fig. 3.8.

Electron-density plots for the filled valence-shell molecular orbitals of these two alkyl halides (CH_3Cl and CH_3F) and two alkyl pseudohalides (CH_3NC and CH_3CN) are shown in Figs. 3.10–3.13 (pp. 36–37) with the related molecular orbitals placed side by side in the same row. The molecular orbitals of methyl fluoride and methyl chloride are given in descending order of stability, with the most stable orbital at the bottom, in Figs. 3.10 and 3.11. However, this is not true for methyl isocyanide or acetonitrile (Figs. 3.12 and 3.13). The reader is referred to Fig. 3.9 for the correct ordering of the valence-shell molecular orbitals of this group of molecules.

CH_3F and CH_3Cl—Molecular orbitals $5a_1$ of methyl chloride and $3a_1$ of methyl fluoride represent the situation in which the valence-shell s atomic orbitals of all of the atoms overlap with each other.

However, the strong electron-withdrawing power of the fluorine atom has garnered much of the electronic charge of this orbital around the fluorine nucleus, thereby noticeably reducing the charge on the carbon and lowering the charge on the hydrogens to an inappreciable value. This can also be seen in Table 3.I, in which the Mulliken gross and overlap populations for the valence orbitals of the molecules represented in Figs. 3.10–3.13 and Figs. 3.15–3.17 are shown.

Orbitals $6a_1$ of methyl chloride and $4a_1$ of methyl fluoride are again very similar, except that the chlorine atom comes from the third period whereas the fluorine atom comes from the second period of the periodic table. According to Table 3.I, the gross population on the chlorine atom $(0.54\ e)$ for orbital $6a_1$ is considerably greater than that on the fluorine atom $(0.31\ e)$ for orbital $4a_1$, and this finding emphasizes the fact that the outer lobe of the chlorine is considerably more diffuse than that of the fluorine. The pair of $2e$ molecular orbitals on methyl chloride and the pair of $1e$ orbitals on methyl fluoride involve C–H and C–X (where X = Cl or F) bonding, with the former being larger in the methyl chloride and the latter being much greater in the methyl fluoride, as can be verified from Table 3.I. This pair of orbitals of E symmetry must exhibit equally filled π character with respect to the C–X bond.

Orbitals $7a_1$ of methyl chloride and $5a_1$ of methyl fluoride are dominated by C–X (p_σ–p_σ) bonding, along with an appreciable amount of halogen lone-pair character. Note the differing polarities of the C–Cl and C–F bonds in the electron-density plots of these molecular orbitals. The pair of $3e$ orbitals on methyl chloride and the pair of $2e$ orbitals on methyl fluoride correspond to the halogen lone-pair electrons, and, particularly in the case of methyl fluoride, this e pair of orbitals also exhibits some C–H bonding. As evidenced by the Mulliken population analysis of Table 3.I, the C–H bond in methyl chloride and methyl fluoride is not expected to exhibit an appreciably different total-overlap population. Therefore, the deficiency in C–H bonding in orbitals $3a_1$ and the $1e$ pair in methyl fluoride as compared to the equivalent orbitals of methyl chloride is made up in the $2e$ pair of methyl fluoride orbitals. Note that the electron densities in Figs. 3.10 and 3.11 ascribed to the pairs of molecular orbitals of E symmetry really correspond to only one orbital of each pair, since the other must exhibit no electron density in the plane chosen for display when this is a nodal plane for the latter orbital.

CH₃NC and CH₃CN—Because the isocyano and cyano groups have one more pair of electrons than

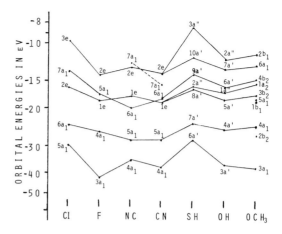

Fig. 3.9. A plot of the orbital energies interrelating the valence-shell molecular orbitals of various CH₃Z, where Z = Cl, F, NC, CN, SH, OH, and OCH₃.

do the halogens, acetonitrile and methyl isocyanide will exhibit one more filled valence-shell molecular orbital than do the methyl halide molecules. Again with the methyl isocyanide and acetonitrile, as shown in Figs. 3.12 and 3.13, the lowest valence-shell molecular orbital is based on valence-shell atomic s character; and a gradual progression in the shape of this orbital can be seen on going from CH₃F to CH₃NC to CH₃CN (also check Table 3.I).

The next molecular orbital in order of decreasing stability for methyl isocyanide and acetonitrile involves s atomic orbitals on the terminal atoms of the CNC or CCN chain of atoms, with a node at right angles to the axis of this three-atom chain being introduced by the use of a p orbital on the central atom of the three. Thus, orbitals $5a_1$ of CH₃NC and CH₃CN exhibit a nodal plane perpendicular to the C_3 axis of the molecule, as do orbitals $4a_1$ of CH₃F and $6a_1$ of CH₃Cl.

Fig. 3.10. Cross-sectional electron-density plots of the valence-shell molecular orbitals of methyl chloride in a plane passing through the carbon, the chlorine, and one of the hydrogen atoms.

Fig. 3.11. Cross-sectional electron-density plots of the valence-shell molecular orbitals of methyl fluoride in a plane passing through the carbon, the fluorine, and one of the hydrogen atoms.

Fig. 3.12. Cross-sectional electron-density plots of the valence-shell molecular orbitals of methyl isocyanide in the plane passing through both carbon atoms, the nitrogen, and one of the hydrogen atoms.

Fig. 3.13. Cross-sectional electron-density plots of the valence-shell molecular orbitals of acetonitrile (methyl cyanide) in the plane passing through both carbon atoms, the nitrogen, and one of the hydrogen atoms.

Fig. 3.10.   CH₃Cl.

Fig. 3.11.   CH₃F.

Fig. 3.12. CH₃NC.

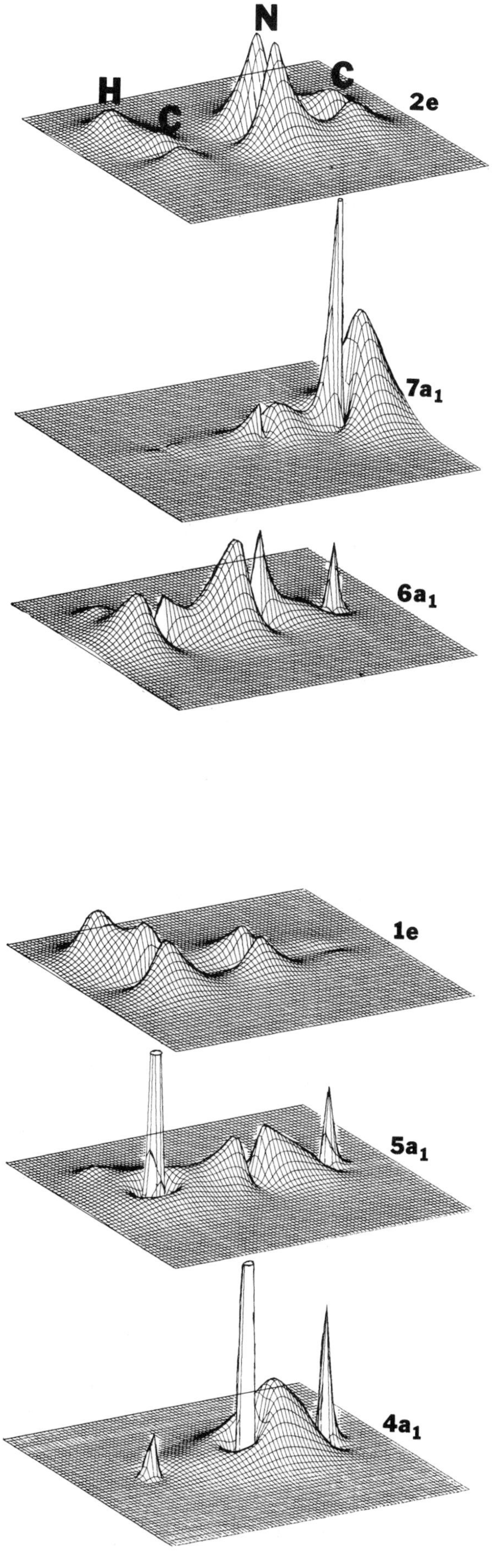

Fig. 3.13. CH₃CN.

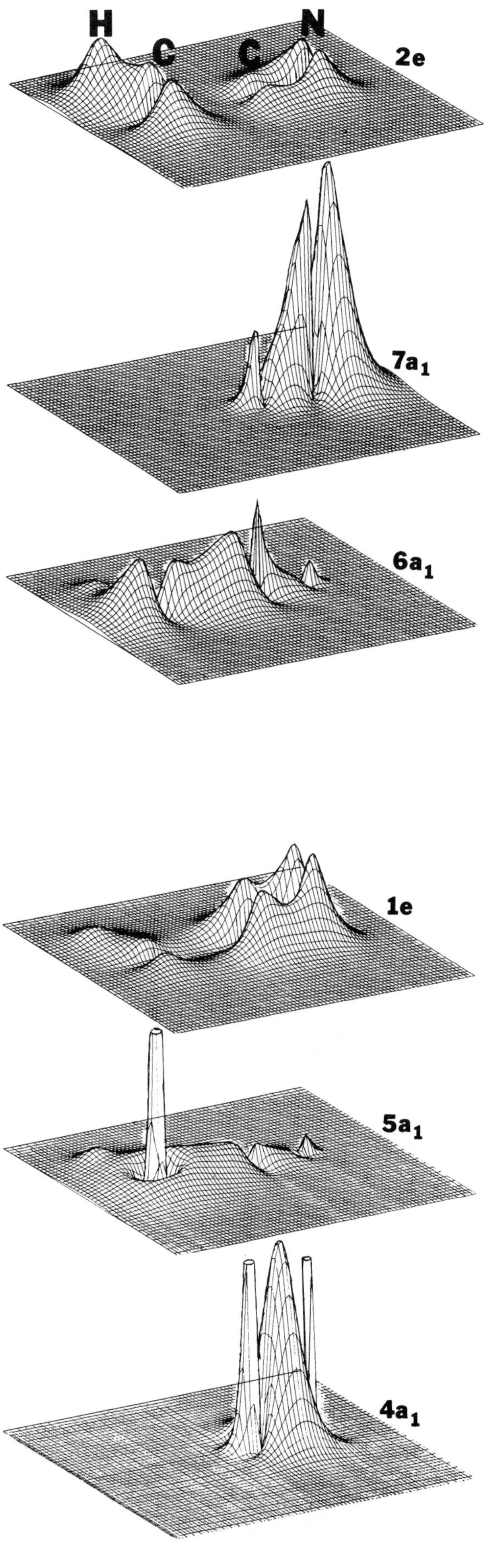

Table 3.I   Mulliken Electronic Populations of Some H₃C—X—Y Molecules

	CH_3Cl		CH_3F		CH_3NC		CH_3CN		CH_3SH		CH_3OH		CH_3OCH_3	
Electronic population	Orb.	Pop.	Orb.	Pop.	Orb.	Pop.	Orb.	Pop.	Orb.	Pop.	Orb.	Pop.	Orb.	Pop.
					$7a_1$	0.04	$7a_1$	−0.01						
C_{gross}									$3a''$	0.02	$2a''$	0.17	$2b_1$	0.81
(methyl C)	$3e$	0.08	$2e$	0.87	$2e$	0.44	$2e$	1.70						
									$10a'$	0.38	$7a'$	0.32	$6a_1$	0.18
	$7a_1$	0.72	$5a_1$	0.43	$6a_1$	0.74	$6a_1$	0.96	$9a'$	0.62	$6a'$	0.82	$4b_2$	0.53
									$2a''$	1.09	$1a''$	0.93	$1a_2$	0.57
	$2e$	2.18	$1e$	1.39	$1e$	1.86	$1e$	0.64						
									$8a'$	0.92	$5a'$	0.58	$3b_2$	0.43
													$5a_1$	0.50
													$1b_1$	0.45
	$6a_1$	0.97	$4a_1$	1.22	$5a_1$	1.03	$5a_1$	1.34	$7a'$	0.66	$4a'$	1.13	$4a_1$	0.56
													$2b_2$	0.66
	$5a_1$	0.49	$3a_1$	0.31	$4a_1$	0.22	$4a_1$	−0.04	$6a'$	0.79	$3a'$	0.32	$3a_1$	0.29
Atomic charge[a]		−0.45		−0.22		−0.33		−0.58		−0.49		−0.28		−0.26
					$7a_1$	0.00	$7a_1$	−0.00						
H_{gross}[b]									$3a''$	0.00	$2a''$	0.00	$2b_1$	0.13
(methyl H)	$3e$	0.09	$2e$	0.40	$2e$	0.17	$2e$	1.44						
									$10a'$	0.01	$7a'$	0.20	$6a_1$	0.07
	$7a_1$	0.04	$5a_1$	0.06	$6a_1$	0.13	$6a_1$	0.10	$9a'$	0.25	$6a'$	0.39	$4b_2$	0.03
									$2a''$	0.00	$1a''$	0.00	$1a_2$	0.22
	$2e$	0.51	$1e$	0.21	$1e$	0.39	$1e$	0.11						
									$8a'$	0.36	$5a'$	0.04	$3b_2$	0.09
													$5a_1$	0.01
													$1b_1$	0.12
	$6a_1$	0.16	$4a_1$	0.15	$5a_1$	0.10	$5a_1$	0.14	$7a'$	0.14	$4a'$	0.17	$4a_1$	0.11
													$2b_2$	0.06
	$5a_1$	0.03	$3a_1$	0.01	$4a_1$	0.01	$4a_1$	−0.00	$6a'$	0.08	$3a'$	0.01	$3a_1$	0.01
Atomic charge[a]		+0.17		+0.17		+0.20		+0.21		+0.18		+0.19		+0.19
					$7a_1$	1.88	$7a_1$	0.31						
X_{gross}									$3a''$	1.88	$2a''$	1.46	$2b_1$	1.31
	$3e$	3.65	$2e$	1.92	$2e$	0.93	$2e$	0.23						
									$10a'$	1.35	$7a'$	1.19	$6a_1$	1.18
	$7a_1$	1.17	$5a_1$	1.39	$6a_1$	0.11	$6a_1$	0.75	$9a'$	0.76	$6a'$	0.72	$4b_2$	0.18
									$2a''$	0.10	$1a''$	0.50	$1a_2$	0.00
	$2e$	0.28	$1e$	2.00	$1e$	0.14	$1e$	1.58						
									$8a'$	0.31	$5a'$	0.90	$3b_2$	0.77
													$5a_1$	0.50
													$1b_1$	
	$6a_1$	0.54	$4a_1$	0.31	$5a_1$	0.23	$5a_1$	0.23	$7a'$	0.83	$4a'$	0.29	$4a_1$	0.29
													$2b_2$	0.25
	$5a_1$	1.41	$3a_1$	1.67	$4a_1$	0.45	$4a_1$	0.63	$6a'$	0.92	$3a'$	1.46	$3a_1$	1.37
Atomic charge[a]		−0.06		−0.28		−0.52		+0.18		−0.14		−0.51		−0.44
					$7a_1$	0.08	$7a_1$	1.70						
Y_{gross}									$3a''$	0.00	$2a''$	0.00	$2b_1$	0.81
					$2e$	2.10	$2e$	0.66						
									$10a'$	0.17	$7a'$	0.08	$6a_1$	0.18
					$6a_1$	0.77	$6a_1$	−0.03	$9a'$	0.38	$6a'$	0.06	$4b_2$	0.53
									$2a''$	0.00	$1a''$	0.00	$1a_2$	0.57
					$1e$	0.82	$1e$	1.44						
									$8a'$	0.13	$5a'$	0.26	$3b_2$	0.43
													$5a_1$	0.50
													$1b_1$	0.45
					$5a_1$	0.44	$5a_1$	0.01	$7a'$	0.14	$4a'$	0.08	$4a_1$	0.56
													$2b_2$	0.66
					$4a_1$	1.30	$4a_1$	1.42	$6a'$	0.07	$3a'$	0.20	$3a_1$	0.29
Atomic charge[a]						+0.24		−0.22		+0.11		+0.22		−0.26

Table 3.I Mulliken Electronic Populations of Some H_3C—X—Y Molecules—(continued)

Electronic population	CH_3Cl Orb.	Pop.	CH_3F Orb.	Pop.	CH_3NC Orb.	Pop.	CH_3CN Orb.	Pop.	CH_3SH Orb.	Pop.	CH_3OH Orb.	Pop.	CH_3OCH_3 Orb.	Pop.
					$7a_1$	0.00	$7a_1$	0.01						
C—H^b-overlap									$3a''$	0.00	$2a''$	0.00	$2b_1$	0.00
(methyl group)	$3e$	0.03	$2e$	0.28	$2e$	0.17	$2e$	0.42						
									$10a'$	0.12	$7a'$	0.16	$6a_1$	0.08
	$7a_1$	0.02	$5a_1$	0.04	$6a_1$	0.15	$6a_1$	0.11	$9a'$	0.00	$6a'$	0.36	$4b_2$	0.28
									$2a''$	0.00	$1a''$	0.00	$1a_2$	0.00
	$2e$	0.46	$1e$	0.21	$1e$	0.34	$1e$	0.10	$8a'$	0.30	$5a'$	0.04	$3b_2$	0.00
													$5a_1$	0.22
													$1b_1$	0.00
	$6a_1$	0.20	$4a_1$	0.20	$5a_1$	0.12	$5a_1$	0.17	$7a'$	0.16	$4a'$	0.21	$4a_1$	0.09
													$2b_2$	0.11
	$5a_1$	0.04	$3a_1$	0.01	$4a_1$	0.01	$4a_1$	0.00	$6a'$	0.09	$3a'$	0.01	$3a_1$	0.01
Total^c		0.74		0.73		0.80		0.80		0.77		0.77		0.79
					$7a_1$	0.00	$7a_1$	−0.06						
C—X overlap									$3a''$	−0.04	$2a''$	−0.15	$2b_1$	−0.09
	$3e$	−0.05	$2e$	−0.35	$2e$	−0.50	$2e$	−0.32						
									$10a'$	0.02	$7a'$	−0.11	$6a_1$	−0.04
	$7a_1$	0.33	$5a_1$	0.19	$6a_1$	0.06	$6a_1$	0.44	$9a'$	0.24	$6a'$	0.18	$4b_2$	0.15
									$2a''$	0.07	$1a''$	0.17	$1a_2$	0.00
	$2e$	0.15	$1e$	0.41	$1e$	0.36	$1e$	0.24	$8a'$	0.12	$5a'$	0.23	$3b_2$	0.17
													$5a_1$	0.10
													$1b_1$	0.13
	$6a_1$	−0.12	$4a_1$	0.02	$5a_1$	0.22	$5a_1$	0.24	$7a'$	−0.13	$4a'$	−0.01	$4a_1$	−0.03
													$2b_2$	0.11
	$5a_1$	0.32	$3a_1$	0.35	$4a_1$	0.18	$4a_1$	−0.06	$6a'$	0.33	$3a'$	0.31	$3a_1$	0.28
Total^c		0.63		0.62		0.32		0.48		0.60		0.62		0.63
					$7a_1$	−0.44	$7a_1$	0.40						
X—Y overlap									$3a''$	0.00	$2a''$	0.00	$2b_1$	0.00
					$2e$	0.76	$2e$	0.41						
									$10a'$	−0.04	$7a'$	0.07	$6a_1$	0.08
					$6a_1$	0.10	$6a_1$	−0.10	$9a'$	0.25	$6a'$	0.07	$4b_2$	0.28
									$2a''$	0.00	$1a''$	0.00	$1a_2$	0.00
					$1e$	0.12	$1e$	0.92	$8a'$	0.06	$5a'$	0.19	$3b_2$	0.00
													$5a_1$	0.22
													$1b_1$	0.00
					$5a_1$	0.18	$5a_1$	−0.01	$7a'$	0.19	$4a'$	0.12	$4a_1$	0.09
													$2b_2$	0.11
					$4a_1$	0.51	$4a_1$	0.68	$6a'$	0.09	$3a'$	0.24	$3a_1$	0.01
Total^c						1.21		2.28		0.54		0.67		0.79

^{a} The atomic charge is obtained by subtracting the total gross population^c of the chosen atom from its atomic number.

^{b} For the CH_3SH(X=S, Y=H), CH_3OH(X=O, Y=H), and CH_3OCH_3(X=O, Y=C) molecules, the chosen hydrogen lies in the C—X—Y plane.

^{c} The total value includes the contributions of all orbitals, including the core orbitals.

As can be seen by comparing Figs. 3.12 and 3.13 with Fig. 3.11, the two pairs of e-type molecular orbitals on the methyl isocyanide and acetonitrile compare well with their equivalent orbitals on methyl fluoride. The additional valence-shell molecular orbital introduced by substituting the isocyano or cyano group for the halogen atom appears in the $7a_1$ plus $6a_1$ set of molecular orbitals of CH_3NC or CH_3CN, both of which are dominated by $2p$ atomic orbitals that overlap end to end. Note that the $6a_1$ orbital of CH_3NC or CH_3CN is very close in electron distribution to the parent CH_3F orbital ($5a_1$). The reader should note that all of the molecular orbitals of methyl isocyanide are more similar to those of methyl fluoride than are the molecular orbitals of acetonitrile. This series, CH_3F, CH_3NC, and CH_3CN, results because the methyl group is successively attached to a second-period atom of groups VII, V, and IV, respectively. From this kind of reasoning one would expect that, except for the difference in overall molecular symmetry, the molecular orbitals of methyl alcohol (where the methyl group is bonded to a group VI atom) would appear even more similar to those of methyl fluoride than do the molecular orbitals of the two methyl pseudohalides; this may be seen by reference to Fig. 3.16. Likewise, there should be a close similarity between the molecular orbitals of methyl chloride and those of methyl mercaptan (compare Figs. 3.10 and 3.15 and the pertinent data of Table 3.I).

D. Methyl Mercaptan, Methyl Alcohol, and Dimethyl Ether

This set of molecules was studied [2] in a Gaussian basis set in which three s-type orbitals were allotted to each hydrogen, five s- and three p-type to the carbon or oxygen atom, and nine s-, five p-, and one d-type to the sulfur, with atom optimization of all exponents except for the molecularly adjusted sulfur d function. For methyl alcohol and methyl mercaptan the staggered form of the molecule is portrayed. The reference planes for the electron-density diagrams pass through the C–O or C–S bond axis and include the hydrogen of the hydroxyl or mercapto group, as well as the trans methyl-group hydrogen. This same general orientation is used for dimethyl ether, with the hydroxyl hydrogen of the methanol being replaced by the second methyl group of the dimethyl ether. The orientation of this second methyl group is set up so that a plane of symmetry passes through the oxygen and bisects the distance between the two methyl groups of the ether.

Electron-density plots are shown in Fig. 3.14 for the total and valence-shell electrons of methyl alcohol and methyl mercaptan. As is the case for the methyl halides in Fig. 3.8, the electrons corresponding to the atoms of the second period (e.g., oxygen) are held more closely to the nucleus than are those of the corresponding third-period atom (sulfur). This effect shows up clearly in the valence-shell plots of Fig. 3.14 since this plot for

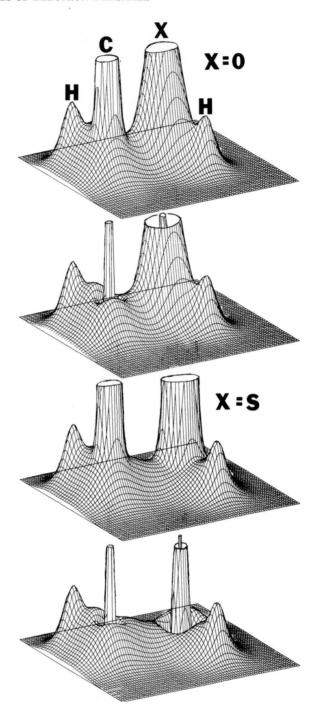

Fig. 3.14. Cross-sectional electron-density plots showing the total and total-valence electronic structure of methyl alcohol (upper two diagrams) and of methyl mercaptan (lower two diagrams) in the plane passing through the *trans* methyl hydrogen, the carbon, and the OH or SH group.

the methyl mercaptan is much flatter than that for the methyl alcohol.

CH_3SH—Electron-density plots for the five out of a total of seven filled valence-shell molecular orbitals of methyl mercaptan that exhibit electron density in a plane passing through the carbon and sulfur atoms, as well as through one of the methyl hydrogens and the hydroxyl hydrogen oriented trans to it, are shown in Fig. 3.15 (p. 42) for this plane. These orbitals for methyl mercaptan are closely related to those of methyl chloride, as may be seen by comparing Fig. 3.15 with Fig. 3.10. The rationalization in terms of bonding character of the SCF valence-shell orbitals of the methyl mercaptan is similar to that given above for the corresponding orbitals of methyl chloride. Thus for CH_3SH, orbital $6a'$ corresponds to the interaction between the valence s orbitals of all the constituent atoms, whereas for orbital $7a'$ this interaction is modified by the inclusion of a nodal plane lying perpendicular to the C–S bond axis.

Orbital $8a'$ is essentially π-like with respect to the C–S bond axis, being similar to orbital $2e$ of CH_3Cl. However, it should be noted that for the CH_3SH molecule the plane of the node passing through the carbon atom is not exactly in line with that of the node passing through the sulfur atom. Orbital $9a'$ exhibits σ-type C–S bonding involving the carbon 2p and the sulfur 3p atomic orbitals. Again, the nodes passing through the carbon and the sulfur atoms are not oriented exactly perpendicular to the C–S bond axis. An even more complicated perturbation of the bonding, in proceeding from methyl chloride to methyl mercaptan, is found in orbital $10a'$ of CH_3SH, an orbital we have formally related to the appropriate $3e$ orbital of CH_3Cl. However, the symmetry of the $10a'$ orbital of CH_3SH is so greatly modified by the geometry of this molecule that this orbital may in no way be considered as exhibiting C–S π-bond character. It does show an interesting interaction between the methyl-group carbon and the mercapto hydrogen, corresponding to a positive Mulliken overlap population of $0.053\ e$. For nonbonded atoms, the overlap population is often negative and, when it is positive, it is generally less than $0.015\ e$.

If it is considered that molecular orbitals $8a'$ and $2a''$ of CH_3SH correspond to the $2e$ orbital pair of CH_3Cl, with CH_3SH orbitals $10a'$ and $3a''$ being related to the $3e$ pair of CH_3Cl, the energy splitting resulting from removing the degeneracy of the C_{3v} symmetry by going to the C_s symmetry of methyl mercaptan may be estimated. Because the calculated energy of CH_3SH molecular orbital $8a'$ is -16 eV as compared to -15.5 eV for orbital $2a''$, the splitting is only 0.5 eV. However, for the more seriously disturbed set of orbitals, $10a'$ at -11.6 eV and $3a''$ at -8.3 eV, the splitting is 3.3 eV, in accord with the observation that the C 2p and S 3p nodal planes in orbital $10a'$ are turned so as to be nearly at right angles to the C–S bond axis.

The differences noted herein between the electron-density distributions of interrelated molecular orbitals of methyl mercaptan and methyl chloride follow, of course, from the fact that these molecules exhibit C_s and C_{3v} symmetries, respectively. Thus, different mixing of orbital characteristics to give an observed SCF molecular orbital is allowed for these different symmetries so that, for example, the $10a'$ orbital of CH_3SH may be considered as being built up from mixing certain CH_3Cl virtual or filled a_1 orbitals into the $3e$ orbital of CH_3Cl on reduction of its C_{3v} symmetry into C_s. Probably a more instructive explanation is simply to attribute the observed differences to the perturbing effect of the mercapto hydrogen on the electronic structure of the atomic array consisting of the methyl group and its associated third-period atom. Comparison of Figs. 3.10–3.13 (pp. 36–37) with Figs. 3.15–3.17 (pp. 42–43) clearly demonstrates this ratiocination.

CH_3OH—In Fig. 3.16 are shown the electron-density plots of the five orbitals appearing in the chosen plane of methyl alcohol, the molecule of which is oriented the same as is the methyl mercaptan in Fig. 3.15. Note the close similarity between the orbitals of methyl alcohol in Fig. 3.16 and the orbitals plotted for methyl fluoride in Fig. 3.11. The presence of the hydroxyl hydrogen in methyl alcohol removes the E-type degeneracy that is associated with the C_{3v} symmetry of methyl fluoride. However, the two orbitals making up the π-like pairs of molecular orbitals of methyl alcohol have rather close-lying energies. Thus, the calculated energy of orbital $5a'$ is -18.2 eV, whereas that of its π-like paired orbital, $1a''$, is -16.7 eV and the energy of orbital $7a'$ is calculated to be -13.0 eV as compared to a value of -11.7 eV for its paired orbital, $2a''$.

The interpretations of the molecular orbitals of methanol are similar to those already given for methyl fluoride, except for the effect of the lower molecular symmetry on the orientation of the 2p oxygen atomic orbitals in molecular orbitals $5a'$ and $7a'$ of the CH_3OH molecule (see the preceding section for a discussion of this matter for the CH_3SH and CH_3Cl molecules). Note that the hump of charge connecting the methyl carbon to the mercapto hydrogen in orbital $10a'$ of CH_3SH is missing from orbital $7a'$ of methyl alcohol. In

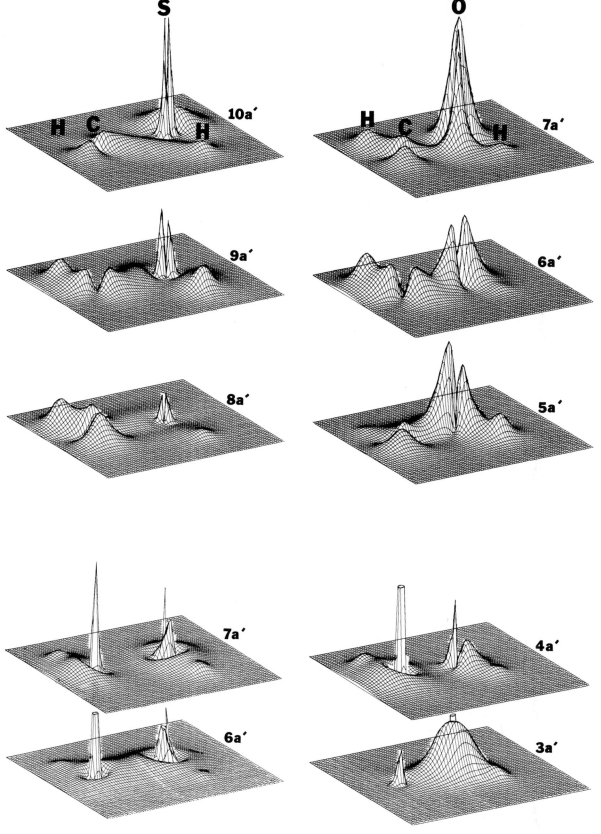

Fig. 3.15. Cross-sectional electron-density plots of the valence-shell molecular orbitals of methyl mercaptan in the plane passing through the *trans* methyl hydrogen, the carbon, the sulfur, and the mercapto hydrogen atom.

Fig. 3.16. Cross-sectional electron-density plots of the valence-shell molecular orbitals of methyl alcohol in the plane passing through the *trans* methyl hydrogen, the carbon, the oxygen, and the hydroxyl hydrogen atom.

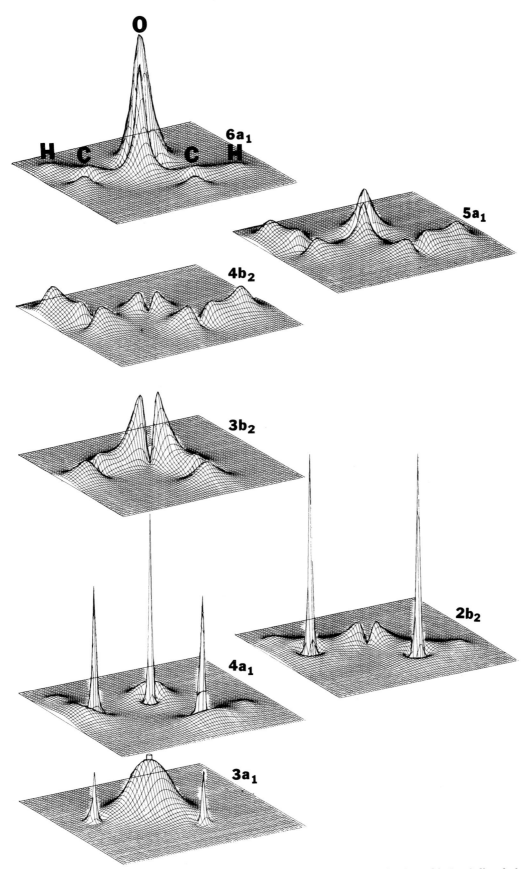

Fig. 3.17. Cross-sectional electron-density plots of the valence-shell molecular orbitals of dimethyl ether in a plane passing through both carbon atoms, the oxygen, and a methyl hydrogen on each of the carbon atoms.

accord with this, the Mulliken overlap population between the methyl carbon and hydroxyl hydrogen of CH_3OH is found to be $-0.008\,e$, corresponding to a "normal" repulsive interaction.

CH_3OCH_3—Not only can the valence orbitals of methyl alcohol be correlated with those of methyl mercaptan (as well as with those of methyl fluoride and methyl chloride) but there is also a close relationship between the valence-shell molecular orbitals of methyl alcohol and those of similarly oriented dimethyl ether. This is apparent from a comparison of Fig. 3.16 with Fig. 3.17. In this comparison, the similarity between orbitals $3a_1$ and $4a_1$ of CH_3OCH_3 and $3a_1$ and $4a_1$, respectively, of CH_3OH are immediately apparent. Not only is this similarity seen for the molecular orbitals based only on s-type functions but it also appears in those molecular orbitals involving p atomic orbitals. Thus, orbitals $3b_2$, $4b_2$, and $6a_1$ of CH_3OCH_3 are very similar to orbitals $5a'$, $6a'$, and $7a'$ of CH_3OH. This similarity also extends to the orbitals exhibiting nodal planes which coincide with the planes of Figs. 3.16 and 3.17, so that orbitals $1a_2$ and $2b_1$ of CH_3OCH_3 correlate with orbitals $1a''$ and $2a''$, respectively, of CH_3OH, orbitals, which in turn correlate with the proper $1e$ and $2e$ orbitals, respectively, of CH_3F. As dimethyl ether exhibits six more electrons than does methyl alcohol, the former molecule must have three more filled valence-shell molecular orbitals and two of these are shown displaced to the right in Fig. 3.17. In addition, there is one more filled molecular orbital, $1b_1$, which has a nodal plane in the plane of this figure and may be interpreted as forming a π-like pair with orbital $5a_1$. Note in Fig. 3.17 that orbital $2b_2$, which is based on the s valence orbitals of the carbon and hydrogen atoms of the methyl group, incorporates a nodal plane in the molecule by use of oxygen p orbitals. Orbital $2b_2$ of dimethyl ether may also be fruitfully related to orbital $5a_1$ of either methyl isocyanide or acetonitrile (see Figs. 3.12 and 3.13). Thus, of the filled molecular orbitals of dimethyl ether utilizing carbon s character, $3a_1$ has no nodal planes, $2b_2$ has one, and $4a_1$ has two, and this is the order of decreasing stability, as shown in Fig. 3.9 (p. 35). Molecular orbital $5a_1$ is obviously closely related to orbitals $4b_2$ and $6a_1$ because it partakes of the characteristics of each.

Fig. 3.18. Cross-sectional electron-density plots showing the total and total-valence electronic structure of lithium fluoride as well as the structure of its four valence orbitals, with all of these plots corresponding to a plane passing through the two atoms of this diatomic molecule.

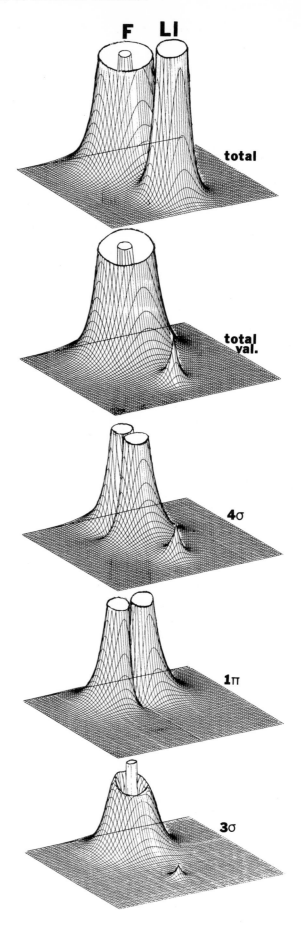

E. Monomeric Lithium Fluoride and Monomeric Methyl Lithium

Although lithium fluoride is generally known in the form of a salt crystal (or a molten or dissolved salt) and lithium methyl exhibits a polymerized structure under normal conditions, it is of academic interest to examine the electronic structure of the gas-phase spectroscopic molecules, LiF and LiCH₃.

LiF—In Fig. 3.18 electron-density plots are presented for a plane passing through the LiF bond axis of diatomic lithium fluoride. The SCF calculations [3] from which these plots were made employed an extended Slater basis set, larger than a double zeta and utilizing d functions on both the lithium and fluorine. The top diagram in this figure, in which the lithium appears in front of the fluorine, corresponds to the total electron density. The valence orbitals are depicted in the next plot and this is followed by representations of the electron-density distribution in three of the four valence-shell molecular orbitals of this molecule. The most stable of these molecular orbitals, 3σ, is dominated by the s valence orbitals of both the lithium and the fluorine. It is clear from the electron-density plot of orbital 3σ that it contributes little if any to the Li–F bonding and that most of the charge contributed to this orbital by the lithium has been transferred to the fluorine. The 1π set of orbitals of lithium fluoride, a set that is solely lone pair, represents a ring of charge around the fluorine atom, with the plane of the ring being perpendicular to the Li–F bond axis. The valence-shell orbital contributing the most to the small amount of Li–F bonding is the least stable one, 4σ. It consists of the end-on interaction of the fluorine p atomic orbital with the lithium s atomic orbital.

For such a highly polar molecule as lithium fluoride, it seems desirable to prepare (a) difference plots between the molecule and its constituent atoms or, similarly, to consider (b) the difference in electron density between the LiF molecule and identically placed Li⁺ and F⁻ ions with the same exponents for the basis functions as those used for the molecule. Such difference plots are shown in Fig. 3.19, at the top of which is a "transparent" diagram indicating that, on molecule formation, the charge in the vicinity of the lithium atom is moved from the region opposite the fluorine to the region close to it. Furthermore, this plot shows that the fluorine exhibits less s character in the molecule than in the atom and that the charge around the fluorine nucleus is also brought from behind the fluorine to the front part facing the lithium atom. Evidence for appreciable

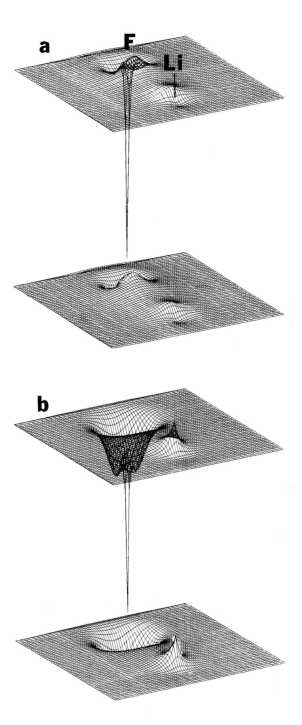

Fig. 3.19. Transparent and regular cross-sectional electron-density plots for all of the electrons of lithium fluoride corresponding to the same cross-sectional plane used for this molecule in Fig. 3.18. Plot (a) corresponds to the molecular electron density minus that obtained by placing the ground-state constituent atoms in the same position in which they appear in the molecule. Plot (b) is a similar one showing the molecular electron density minus that of the Li⁺F⁻ ion pair. The top plot under either (a) or (b) is transparent and therefore projects below the chosen plane.

bonding in this molecule is shown in the second diagram of Fig. 3.19, which is identical to the upper diagram except that it is not in the form of a transparent basket.

The lower two diagrams of Fig. 3.19 both represent the shift in electron density on going from the ions to the molecules. The sharp negative peak in the transparent diagram shows that the fluorine s character in its ion is about the same as

Fig. 3.20. Cross-sectional electron-density plots showing the total and total-valence electronic structure of methyl lithium as well as the structure of its four valence orbitals. All of these plots correspond to a plane passing through the lithium, the carbon, and one of the hydrogen atoms.

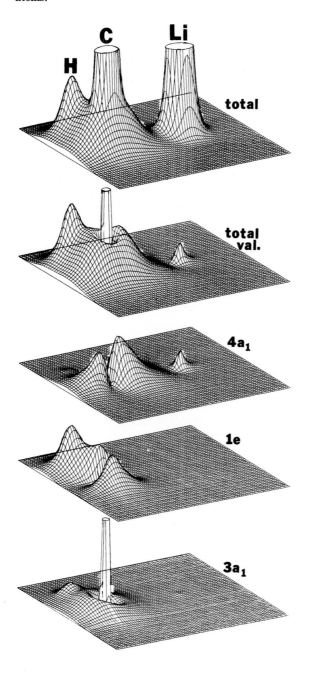

in the atom, a value which is greater than the s character in the LiF molecule. The big negative blob surrounding the fluorine nucleus and the sharply pointed positive peak centered on the lithium nucleus give evidence that an approximation of the LiF molecule as an unpolarized Li^+F^- ion pair is a poor representation because it takes too much charge away from the lithium and puts too much on the fluorine. From Fig. 3.19, it appears that the approximation of the lithium fluoride diatomic molecule as an overlay of its constituent atoms gives a better fit to reality than the representation of this molecule as a pair of unpolarized ions, because considerably more charge is transferred from the lithium to the fluorine in the case of complete ionization than is found for the molecule.

$LiCH_3$—Monomeric methyl lithium was calculated [2] using atom-optimized Gaussian functions in a (53/53/3) basis set. The electron-density diagrams of Fig. 3.20 correspond to a molecular cross section passing through the lithium, the carbon, and one of the hydrogen atoms. The total electron density is shown at the top of this figure, with the electron density of the total valence-shell orbitals beneath it. It is clear from these two diagrams that the methyl group in methyl lithium is quite normal but that the Li–C bond is extremely polar, with the negative end, of course, being at the carbon. Orbitals $3a_1$, $1e$, and $4a_1$ of methyl lithium are closely related to orbitals 3σ, 1π, and 4σ, respectively, of lithium fluoride. For methyl lithium, the overlap between the lithium and carbon atoms was calculated to be 0.56 e, of which 0.43 e is attributable to orbital $4a_1$. According to the Mulliken population analysis, the charge on the carbon atom of -0.87 e is balanced by the charge of $+0.49$ e on the lithium and $+0.13$ e on each hydrogen.

F. Hydrogen Cyanide and Its Analog, Methinophosphide

Although the HCN molecule has been known for a long time, its phosphorus analog, HCP, was first prepared only in 1961; its geometry was investigated shortly thereafter by microwave spectroscopy. Because these two molecules are linear, the σ, π notation applies precisely to them and the electron density for any plane passing through the molecular axis is identical to that obtained for any other plane which includes this axis. Electron-density plots for these molecules are shown in Figs. 3.21 and 3.22, which start with the total electron density followed by the electron density of the sum of the valence orbitals. The calculations

[4] on both molecules were carried out using Gaussian functions, with a (3/52/951) basis set for the HCP and a (3/52/52) for the HCN.

Fig. 3.21. Cross-sectional electron-density plots of the total and total-valence electronic structure of hydrogen cyanide, as well as the structure of its valence orbitals. The basal plane of these plots passes through the axis of this linear molecule.

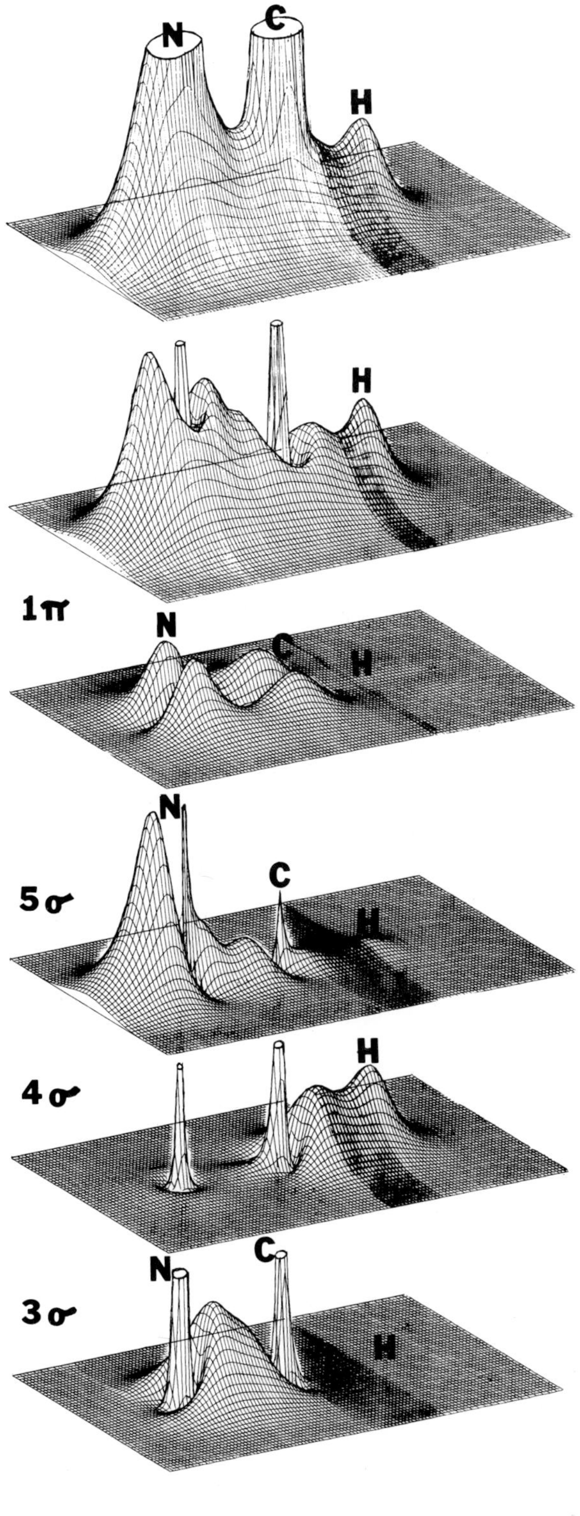

Although the HCP molecule was studied in several basis sets, with and without d orbitals, the shapes of any one of the electron-density plots were found to be affected so little by the change in basis set that it would only be apparent from very detailed observation.

Fig. 3.22. Cross-sectional electron-density plots of the total and total-valence electronic structure of methino phosphide, as well as the structure of its valence orbitals. The basal plane of these plots passes through the axis of this linear molecule.

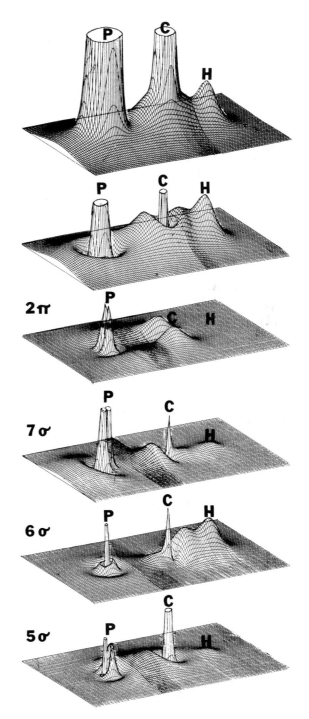

As can be seen from the total electron densities, the electrons in the phosphorus end of the HCP molecule are considerably more spread out than at the nitrogen end of the HCN molecule, in the sense that the phosphorus atom corresponds to a mighty column of charge surrounded by a diffuse charge distribution, whereas the charge density around the nitrogen atom slopes down gradually. In the total-electron-density diagrams of Figs. 3.21 and 3.22, it is apparent that the electron density in the region toward the middle of the P–C bond is considerably lower than that in the corresponding region of the N–C bond. However, the overlap population for the P–C bond was found to be 1.82 e for the HCP calculation and 1.62 e for the N–C bond in HCN. The apparent contradiction between the total electron-density plots and the overlap population analysis is readily explained by the diffuseness of the phosphorus valence electrons as compared to those of the nitrogen.

Comparison of the electron-density plots of the filled valence orbitals of HCN with those of HCP is particularly illuminating. As expected, the H–C end of the molecule looks the same in both cases but there is a great difference at the other end. First of all, the outermost ring of electronic charge around the phosphorus atom, the charge involved in bonding, corresponds to a broad and quite diffuse distribution, as compared to the more closely held ring of charge exhibiting the same role for the nitrogen atom. This means that the phosphorus lone-pair electrons are obviously quite diffuse as compared to the closely held nitrogen lone-pair electrons. From the usual electronegativity values, nitrogen would be expected to be electron-withdrawing from carbon, and carbon ought to be electron-withdrawing from phosphorus. These expectations are borne out by the slopes of the electron-density surfaces in the central region of the respective bonds. Hydrogen and phosphorus are supposed to have about the same electronegativities and it can be seen in Fig. 3.22 that the two sides of the lips surrounding the carbon atoms are about the same along the molecular axis in the direction of the phosphorus and of the hydrogen. Conversely, these lips in Fig. 3.21 differ in the way that is expected from the idea that the electronegativity of carbon is about midway between that of hydrogen and nitrogen.

The similarity between the five valence orbitals of these two molecules is striking, as can be seen from Figs. 3.21 and 3.22. It is obvious that the most stable of the valence orbitals (3σ for HCN and 5σ for HCP) involves only the s orbitals and

corresponds primarily to N–C and P–C σ bonding [i.e., (s_σ–s_σ) bonds], respectively. Comparison of the 5σ molecular orbital of HCP with the 3σ of HCN shows the opposing polarities of the P–C and N–C σ bonds. Again, it should be noted that there is a much greater density of electrons on the line connecting the nitrogen and carbon nuclei in the 3σ orbital of HCN than in the line connecting the phosphorus and carbon nuclei in the 5σ orbital of HCP, even though the N–C overlap was calculated to be 0.67 e and the P–C overlap to be 0.63 for these respective orbitals. The C–H overlap population appearing in the 5σ orbital of HCP may be rationalized as resulting from much greater diffuseness of the P–C σ bond as compared to the N–C σ bond, which is not associated with C–H overlap. We can argue that the diffuseness of the P–C bond is so great that it gives an appreciable electronic charge in the neighborhood of the hydrogen nucleus, which will, of course, concentrate this negative charge around itself and in the region between it and the carbon to give an appreciable C–H overlap and H gross population in orbital 3σ.

Orbitals 4σ of HCN and 6σ of HCP are also dominated by s character and correspond primarily to H–C bonding. It is clear from their electron-density plots that the H–C bonding character of these related orbitals is essentially unaffected by the substitution of nitrogen by phosphorus. However, the H–C overlap population for orbital 4σ of HCN was calculated to be 0.70 e, as compared to a value of 0.58 e for 6σ of HCP. Orbitals 5σ of HCN and 7σ of HCP involve carbon 2p with nitrogen 2p or phosphorus 3p atomic orbitals and are dominated by the lone-pair character of the nitrogen and phosphorus atoms, respectively. The Mulliken gross populations for the phosphorus and nitrogen atoms in these orbitals was found to be 1.74 and 1.56 e, respectively. However, the electron-density plots show that the phosphorus lone-pair charge is much more diffuse than that of the nitrogen, with some of the phosphorus 7σ valence-orbital charge being held in the core-region 3p antinodes of this atom. The least stable of these molecular orbitals is the degenerate pair of 1π orbitals on HCN and 2π orbitals on HCP. (Note that the 1π pair of molecular orbitals of HCP corresponds to the pair of phosphorus 2p orbitals in its atomic core, with the lobes of these orbitals lying perpendicular to the molecular axis.) Obviously the pair of 1π orbitals of HCN and of 2π orbitals of HCP correspond to the N–C or the P–C triple bond, respectively. Since the phosphorus atom utilizes its 3p atomic orbitals in the formation of this triple bond, Fig. 3.22 shows "dead storage" of part of these electrons in the inner 3p antinodes, while only the outer antinodes are involved in the bonding. Thus, the two lobes of the P–C π bond do not lie parallel to each other but are cocked to form somewhat of a vee, with the wider part at the phosphorus. Again, the electron density of the N–C π-bonding lobes in the region half way between the bonded atoms is obviously higher than in the same region for the P–C lobes, although the overlap populations are nearly identical: 0.50 e for the pair of 1π orbitals for HCN and 0.51 e for the 2π pair of HCP.

It is interesting to note that disallowing d character to the phosphorus has a significant effect on only one of the overlap populations; i.e., it causes a decrease in the P–C overlap in orbital 7σ of the HCP molecule from a small positive value to a negative one (ca. -0.3 e). With respect to the gross population, the major effect is found for the phosphorus and carbon atomic charges in the case of orbital 5σ, with the disallowing of d character shifting electrons from the phosphorus to the carbon.

The appearance of these orbital-density plots is in accord with the magnitudes of their orbital energies in methinophosphide and hydrogen cyanide. For each of the compared valence orbitals, the methinophosphide molecule in a (951/52/3) basis consistently exhibits higher orbital energies than does hydrogen cyanide in a (52/52/3) basis. Thus, the difference in energy between the N–C and P–C π orbitals is 3.4 eV. For the molecular orbitals dominated by the lone pair, the difference is less, corresponding to 1.8 eV; and for the predominantly C–H σ-bond orbitals, the difference is 2.2 eV. The greatest difference, 8.4 eV, is found between the orbitals dominated by the C–N and C–P σ bonds.

G. Ammonia, Phosphine, and Arsine

Correlation between molecular orbitals should be particularly easy to carry out in a series of molecules in which one of the atoms is substituted by increasingly larger atoms from the same group of the periodic table. Such a series is found in the three molecules ammonia, phosphine, and arsine. These have the formula MH_3, with M standing for N, P, and As, respectively.

Electron-density plots are shown in Figs. 3.23–3.25 for the total molecule as well as for the individual filled valence-shell orbitals of ammonia, phosphine, and arsine, respectively. The cross section represented in Figs. 3.23–3.25 corresponds to the plane passing through the C_3 axis of the molecule (i.e., through the N, P, or As and its unshared pair of electrons) and one of the hydrogen atoms. The calculation for ammonia was carried out in a (52/3) Gaussian basis set; for phosphine in a (951/3) basis set; and for arsine in a (1383/3) basis set [5].

The top diagrams of Figs. 3.23–3.25 show the total electron density in the chosen plane. It is apparent from these figures that the total electron density in the region of the midpoint of the M–H bond is considerably greater for ammonia than for phosphine, which in turn exhibits about the same value as arsine. However, the M–H overlap populations obtained for these particular calculations are closer to each other than might be inferred from the electron-density diagrams, being equal to 0.70 e for NH_3, 0.57 e for PH_3, and 0.66 e for AsH_3. Now, turning to the bottom diagrams in Figs. 3.23–3.25 we see that the most stable valence-shell molecular orbital of each of these MH_3 molecules involves overlap between the s orbital on the M atom and the s orbitals of the three hydrogen atoms. Note that it is the outer antinode of a 2s orbital that is involved in the N–H bonding of the $2a_1$ molecular orbital of ammonia, whereas for the P–H bonding of the $4a_1$ orbital of phosphine and the As–H bonding of the $7a_1$ orbital of arsine, it is the outer antinode of a 3s and 4s orbital, respectively.

The next pair of orbitals of each of these molecules, as measured in order of decreasing stability, is degenerate and exhibits E symmetry. The plot of this pair of $1e$ molecular orbitals for the ammonia molecule again shows a high electron density in the midpoint of the bond formed by the essentially end-on overlap of the nitrogen 2p lobes with the hydrogen 1s orbital [i.e., (p_σ–s_σ) bonding] as compared to the situation for the similar $2e$ and $5e$ pair of molecular orbitals of phosphine and arsine, respectively. Again, the M–H overlap populations of $+0.49$, $+0.59$, and $+0.47$ e per bond for these calculations on NH_3, PH_3, and AsH_3, respectively, give a considerably smaller relative value for the $1e$ orbital of ammonia than would be naïvely inferred from comparison of Figs. 3.23–3.25. This valence-shell pair of e orbitals for each of the molecules corresponds to a ring of charge around the central atom interacting sideways with the three hydrogen atoms.

The least stable of the molecular orbitals ($3a_1$ for NH_3, $5a_1$ for PH_3, and $8a_1$ for AsH_3) is dominated by the lone-pair electrons on the central atom. Note that the lone pair for the nitrogen is held quite close to the atomic nucleus and drops off rapidly with increasing distance away from it along the C_3 axis on the opposite side from the

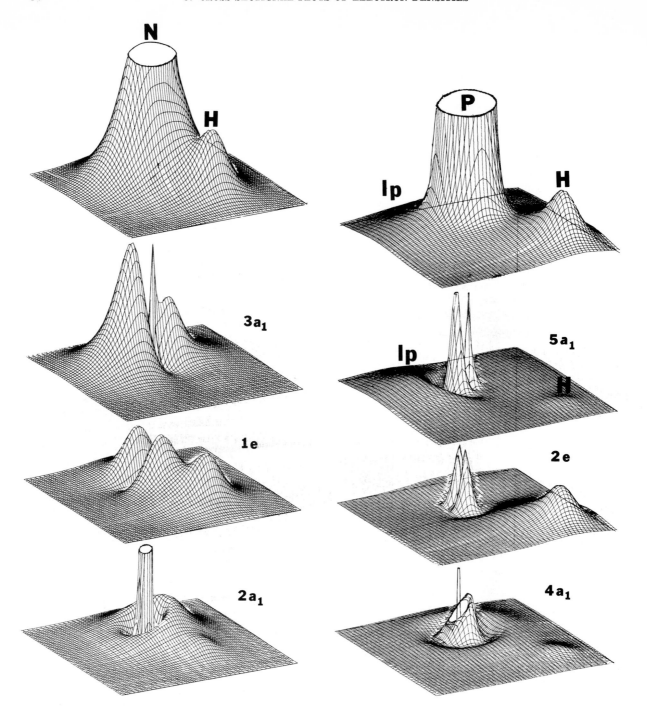

Fig. 3.23. Cross-sectional electron-density plots for the total molecule and the valence-shell orbitals of ammonia for a plane passing through the nitrogen, one of the hydrogen atoms, and the C_3 axis of the molecule.

Fig. 3.24. Cross-sectional electron-density plots for the total molecule and valence-shell orbitals of phosphine for a plane passing through the phosphorus, one of the hydrogen atoms, and the C_3 axis of the molecule.

hydrogen substitutents. For phosphine and arsine the lone pairs are diffuse.

There are two lessons to be learned from the electron-density plots in Figs. 3.23–3.25. First, with respect to the distribution of the electrons available for bonding, there is a much greater

difference between nitrogen and phosphorus than between phosphorus and arsenic. Thus, the bonding and lone-pair regions of the valence orbitals shown in this series of diagrams are more similar for PH_3 and AsH_3 than for NH_3. This comes about because of the need to distribute the nodes

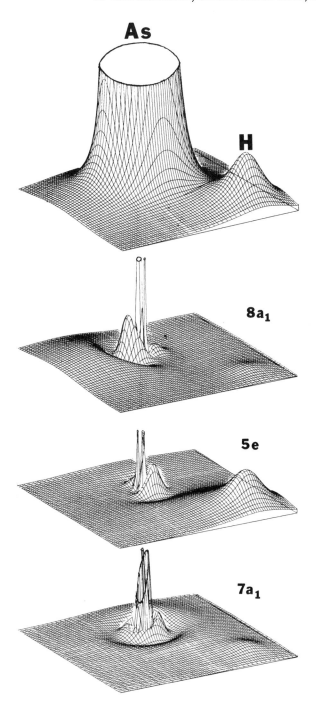

Fig. 3.25. Cross-sectional electron-density plots for the total molecule and valence-shell orbitals of arsine for a plane passing through the arsenic, one of the hydrogen atoms, and the C_3 axis of the molecule.

is 0.09 Å, whereas for phosphorus and arsenic, it is 0.22 and 0.29 Å, respectively [6]. Likewise, the inert gases also illustrate that the core radii [6] of the third- and fourth-period elements are close together, viz. Ne core, 0.055 Å; Ar core, 0.179 Å; Kr core, 0.252 Å; and Xe core, 0.373 Å, showing the following differences (Ar − Ne) = 0.124 Å; (Kr − Ar) = 0.073 Å, (Xe − Kr) = 0.121 Å. The second lesson is that the relative change from molecule to molecule in the Mulliken overlap populations, which correspond to the integral over all space of the chosen atomic orbitals of the particular pair of atoms in question, does not necessarily parallel the change in the electron density directly at the center of the respective bond. Not surprisingly, the calculated variation in either of these properties does not parallel the change in M–H bond energy calculated from heats of formation of the MH_3 molecules at 0°K, viz., 92.3 kcal/mole for N–H, 75.6 for P–H, and 69.7 for As–H.

H. Methylamine, Methylphosphine, and Its Isomeric Phosphorus Ylide

The cross-sectional electron-density graphs for methylamine [2], CH_3NH_2; methylphosphine [7], CH_3PH_2; and methylenephosphorane [8] (the ylide), CH_2PH_3, were obtained from wave functions employing three s-type atom-optimized Gaussian functions to describe each hydrogen atom; five s- and two p-type for the carbon and for the nitrogen; and nine s-, five p-, and one d-type for the phosphorus. The electron-density diagrams shown in Fig. 3.26 for methylamine and Fig. 3.27 for methylphosphine correspond to those filled valence-shell molecular orbitals that exhibit electron density in the mirror plane of the staggered forms of these molecules, the plane that bisects the line connecting the two hydrogens of the NH_2 or PH_2 group and therefore passes through one of the methyl hydrogens. (For both methylamine and methylphosphine there are two valence-shell orbitals for which the planes of Figs. 3.26 and 3.27 are nodal planes. These orbitals of A'' symmetry are not shown here.)

As is the usual case, the most stable orbital in either of these molecules involves overlap between the s atomic orbitals of all of the atoms with their nearest neighbors. Note that molecular orbital $3a'$ of CH_3NH_2 shows the greatest electron density on the nitrogen atom, whereas the equivalent orbital $6a'$ of CH_3PH_2 shows the greatest density on the carbon atom. (The Mulliken population analysis for these calculations indicates a gross

of the orbitals in a relatively restricted volume of space (the volume that corresponds more or less to the core electrons) in order that all of the orbitals of each molecule fulfill the requirement of being mutually orthogonal (i.e., having no net overlap). For the nitrogen atom, the core radius

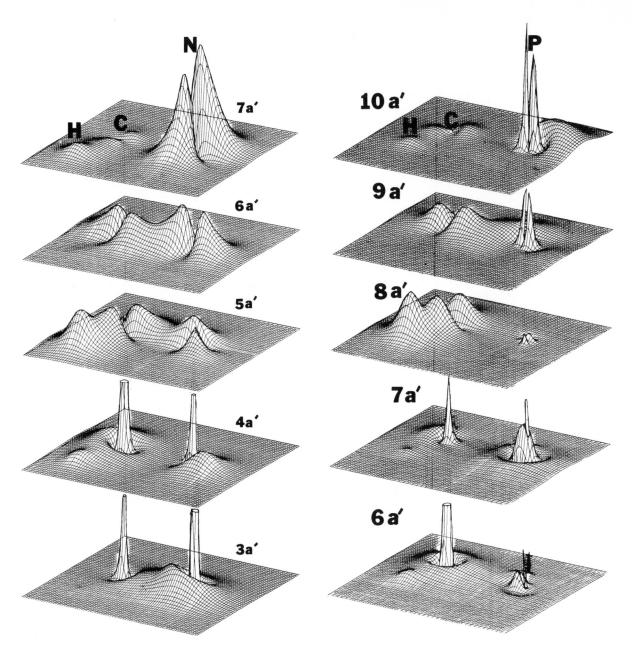

Fig. 3.26. Cross-sectional electron-density plots for the valence-shell orbitals of A′ symmetry of methyl amine. The basal plane of this plot passes through a methyl hydrogen atom, the carbon atom, the nitrogen atom, and the lone-pair axis of the latter.

Fig. 3.27. Cross-sectional electron-density plots for the valence-shell orbitals of A′ symmetry of methyl phosphine. The basal plane of this plot passes through a methyl hydrogen atom, the carbon atom, the phosphorus atom, and the lone-pair axis of the latter.

population of 0.43 e on the C and 1.22 on the N for orbital 3a′ of CH_3NH_2, with 1.16 e on the C and 0.40 on the P for orbital 6a′ of CH_3PH_2.) This behavior is what would be expected from the fact that the electronegativity of carbon lies halfway between that of nitrogen and phosphorus. The next higher line of orbitals (4a′ for CH_3NH_2 and 7a′ for CH_3PH_2) corresponds to antibonding in the C–N and C–P bonds, respectively. The follow-

ing orbitals in decreasing stability are 5a′ for CH_3NH_2 and 8a′ for CH_3PH_2, both of which exhibit π character with respect to the C–N or C–P bond axis, whereas the next ones (6a′ for CH_3NH_2 and 9a′ for CH_3PH_2) correspond to C–N (p_σ–p_σ) and C–P (p_σ–p_σ) bonding, respectively.

The least stable of the molecular orbitals, which appear at the top of Figs. 3.26 and 3.27, are

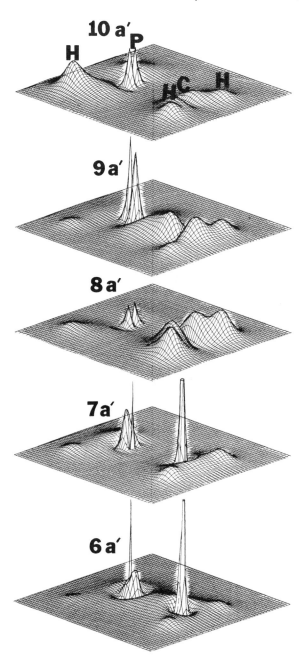

Fig. 3.28. Cross-sectional electron-density plots for the valence-shell molecular orbitals of methylenephosphorane, with the basal plane of the plots passing through the phosphorus atom, one of its hydrogen atoms, and the carbon and two hydrogen atoms of the methylene group.

eclipsed hydrogen on the phosphino group. Two of the orbitals of this molecule exhibit nodal planes that coincide with the basal plane of Fig. 3.28. These orbitals are shown in Fig. 3.29 for the plane

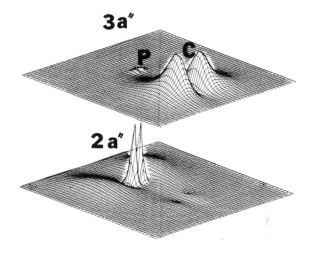

Fig. 3.29. Cross-sectional electron-density plots of the two valence-shell molecular orbitals of methylenephosphorane appearing in a plane passing through the phosphorus and the carbon atom perpendicular to the plane of Fig. 3.28.

passing through the C–P axis at right angles to the plane of the previous figure. Since the hydrogen atoms in methylphosphine, CH_3PH_2, are arranged differently than in methylenephosphorane, CH_2PH_3, the orbitals are not directly comparable. However, when Fig. 3.27 is compared with Fig. 3.28, it can be seen that, for both of these molecules, the two more stable of the valence-shell molecular orbitals involve overlap of the s-type atomic functions of each of the atoms with their nearest neighbor, with C–P bonding for the more stable of this pair and antibonding for the less stable. Likewise, the remaining orbitals of A′ symmetry correspond to $(p_\pi–p_\pi)$, $(p_\sigma–p_\sigma)$, and $(p_\pi–p_\pi)^*$ P–C bonding, respectively, for both molecules. The orbitals of A″ symmetry shown in Fig. 3.29 (i.e., molecular orbitals 2a″ and 3a″ of methylenephosphorane) are dominated by phosphorus lone-pair and carbon lone-pair character, respectively.

The sorting of the A′ and A″ orbitals with respect to energy is somewhat different for the two molecules. Thus, for methylphosphine, the ordering for the valence-shell orbitals from the most to the least stable is 6a′, 7a′, 2a″, 8a′, 3a″, 9a′, and 10a′, whereas for methylenephosphorane it is 6a′, 7a′, 8a′, 2a″, 9a′, 10a′, and 3a″.

dominated by the phosphorus and nitrogen lone pairs. Comparison of these two figures demonstrates the greater diffuseness of the phosphorus lone-pair electrons as compared to those of the nitrogen atom.

Figure 3.28 shows the electron densities of the eclipsed form of methylenephosphorane in the plane passing through the C–P bond as well as the pair of methylene hydrogen atoms and the

I. Cyclopropane, Phosphirane, and Thiirane

Small-ring molecules have long been of interest, primarily because many of them involve angular strain and therefore exhibit enhanced reactivity. From the viewpoint of electronic structure, these cyclic molecules may possibly exhibit the presence of transannular electronic interactions. The smallest cyclic hydrocarbon is cyclopropane, which has many analogs. For example, one of the methylene groups of the cyclopropane may be substituted by an N–H or a P–H group to give azirane (ethylene imine) or phosphirane; or it may be substituted by an oxygen or a sulfur to give oxirane (ethylene oxide) or thiirane. Rather than obtaining wave functions for all five of these molecules in order to elucidate the similarities between their molecular orbitals, only the following three were chosen for study because it appeared obvious that, if their orbitals could be interrelated, there should be no problem of also including the orbitals of azirane and oxirane in the correlation. In these computations [9, 10], a (52) Gaussian description of carbon was employed with a (2) or (3) basis for each hydrogen and a (951) for either phosphorus or sulfur.

$$H_2C \overline{\quad\quad} CH_2$$
$$C$$
$$H_2$$
Cyclopropane

$$H_2C \overline{\quad\quad} CH_2$$
$$P$$
$$H$$
Phosphirane

$$H_2C \overline{\quad\quad} CH_2$$
$$S$$
Thiirane

Electron-density plots corresponding to the ring plane are shown for those filled valence-shell molecular orbitals exhibiting electron density in this plane for cyclopropane, phosphirane, and thiirane in Figs. 3.30–3.32. Comparison of these three figures shows that there is indeed close orbital-to-orbital correlation across this set of cyclic molecules. The most stable of the valence-shell molecular orbitals ($2a_1'$ for cyclopropane, $6a'$ for phosphirane, and $5a_1$ for thiirane) corresponds, as expected, to the interaction of the valence-shell s atomic orbitals of the ring atoms with each other and also to a considerably lesser extent with the methylene hydrogen atoms. From the chemical bond viewpoint, this set of orbitals corresponds to a three-center bond connecting the three ring atoms through a high overlap in the center of the rings.

The next pair of orbitals is degenerate (i.e., has exactly the same energies) in the case of the highly symmetrical (D_{3h}) cyclopropane (the $2e'$ pair) but necessarily differs in energy for the phosphirane and thiirane because of the lower overall symmetry of these molecules. Because the symmetry of thiirane (C_{2v}) is higher than that of

Fig. 3.30. Cross-sectional electron-density plots of the valence-shell molecular orbitals of cyclopropane in the ring plane.

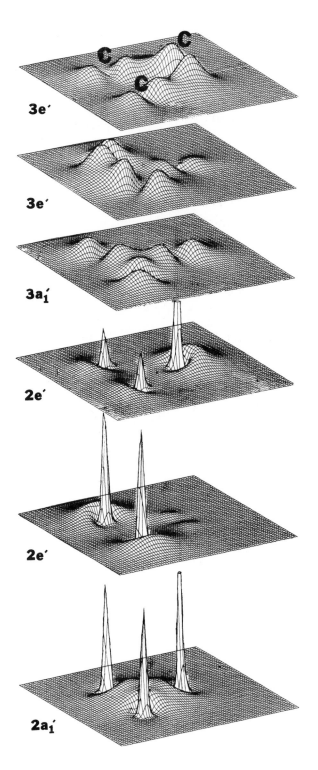

Fig. 3.31. Cross-sectional electron-density plots of the valence-shell molecular orbitals of phosphirane in the ring plane.

Fig. 3.32. Cross-sectional electron-density plots of the valence-shell molecular orbitals of thiirane in the ring plane.

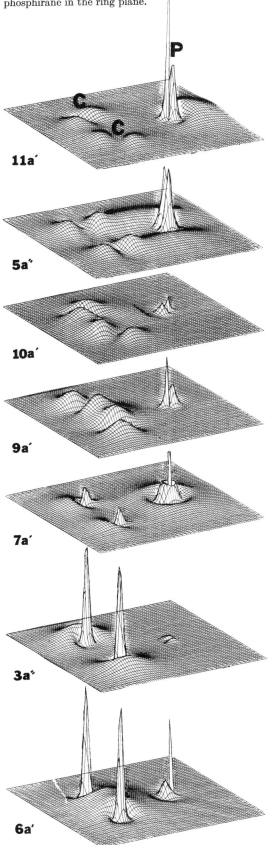

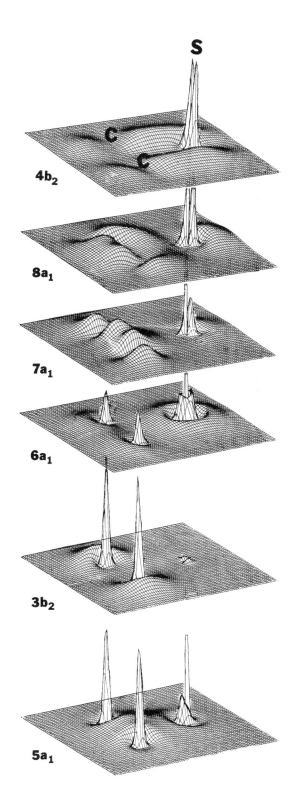

phosphirane (C_s), it would be expected that the energy difference between these pairs of orbitals would be greater for phosphirane than for thiirane. This is found to be the case; since, for thiirane, orbital $3b_2$ exhibits an energy of -22.8 eV and orbital $6a_1$, -22.4 eV, as compared to -22.4 eV for orbital $3a''$ and -20.7 eV for $7a'$ of phosphirane. These pairs of molecular orbitals ($2e'$ for C_3H_6, $7a'$ and $9a'$ for C_2H_4PH, and $6a_1$ and $7a_1$ for C_2H_4S) involve mostly the valence s orbitals of the ring atoms with net antibonding between them. However, they do exhibit some bonding with the hydrogen atoms.

The $3a_1'$ orbital of cyclopropane is quite similar to the $2a_1'$ orbital of this molecule in that the bonds from the ring atoms are again directed toward the center of the ring. However, in the case of molecular orbital $3a_1'$, it is the p atomic orbitals that are involved. In our calculation, the C–C overlap (0.09 e) for orbital $3a_1'$ of cyclopropane is about one-third of that for orbital $2a_1'$ (0.28 e), and this is quite apparent from Fig. 3.30. The nature of the bonding in molecular orbitals $9a'$ of phosphirane and $7a_1$ of thiirane is the same as that for $3a_1'$ of cyclopropane inasmuch as each of the ring atoms have a p lobe directed more or less toward the center of the ring. However, the 2p lobes of the two methylene carbons in each of these molecules are effectively pushed away from the center of the ring by the outermost 3p lobe of the phosphorus or sulfur atom—the lobe that is involved in the bonding. This situation has the effect of increasing the C–C overlap for these orbitals in phosphirane (0.14 e) and thiirane (0.15 e) as compared to cyclopropane (0.10 e). It also leads to a concomitant increase in C–H overlap for phosphirane and thiirane as compared to that for cyclopropane.

The $3e'$ set of degenerate orbitals in cyclopropane corresponds to a positive net C–C overlap, as calculated in various (sp) basis sets. However, when d functions were allowed to the three carbon atoms, it was found that the overlap became quite large and negative (-0.4 e). Whether or not this pair of $3e'$ orbitals is bonding or antibonding, their net effect is to deploy along the C–C bonds (which make up the ring) electron densities, which in a hybrid atomic orbital would be equivalent to lobes directed toward each other from each carbon atom. This is illustrated in Fig. 3.30. Again, because of the lower symmetry of phosphirane and thiirane, the orbital degeneracy is lost. As expected, however, the energies of orbitals $8a_1$ and $4b_2$ of thiirane (-11.3 and -10.5 eV,

respectively) are closer than the energies of orbitals $10a'$ and $5a''$ of phosphirane (-12.0 and -9.7 eV, respectively).

Of the total of nine filled valence-shell molecular orbitals corresponding to each of these molecules, only six exhibit electron density in the ring plane for the cyclopropane and thiirane molecule. Because of the lower symmetry of the phosphirane molecule, seven orbitals exhibit electron density in the C_2H_4PH ring plane. This seventh orbital ($11a'$) is like orbital $10a'$ of phosphirane with an extra nodal plane lying between the C_2H_4 and PH moieties. This orbital also has a considerable hump of electron density showing up in the ring plane behind the phosphorus atom. This orbital is therefore bonding with respect to the C–C and P–H bonds and antibonding with respect to the C–P bonds; it also shows some phosphorus lone-pair character.

The orbitals missing from Figs. 3.30–3.32 are all primarily involved in the bonding between the ring atoms and the peripheral hydrogen atoms. For cyclopropane, these missing orbitals are $1a_2''$, and $1e''$; for thiirane, they are $2b_1$, $1a_2$, and $3b_1$, and for phosphirane, $8a'$ and $4a''$.

An interesting way of studying [9] the effect of ring closure for strained cyclic structures involves the use of electron-density plots. Such an investigation has been carried out for cyclopropane by making a difference map (shown in Fig. 3.33) to represent the density of cyclopropane minus that of a butane molecule which

Fig. 3.33. Cross-sectional electron-density difference plots showing a H_2C—CH_2 edge of cyclopropane minus the superimposed middle CH_2—CH_2 segment of the butane molecule. This plot, which is shown at a fivefold greater electron-density scale than that used in Fig. 3.30, demonstrates the effect of ring closure on the electron-density distribution.

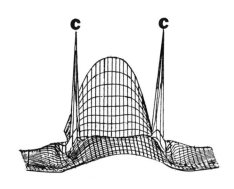

is somewhat distorted so that its central C$_2$H$_4$ group is exactly superimposed on the C$_2$H$_4$ group making up one edge of the cyclopropane ring. This difference plot thus corresponds to the following ring-closure reaction:

$$\text{H}_3\text{C} \quad \text{CH}_3 \atop \text{H}_2\text{C}-\text{CH}_2 \longrightarrow {\text{H}_2\text{C}-\overset{\text{H}_2}{\text{C}}-\text{CH}_2} + \text{CH}_4$$

The plot of Fig. 3.33, which corresponds to the ring plane as viewed from a CH$_2$–CH$_2$ edge looking in toward the center of the ring, shows that ring closure has the following effects: (1) the s electron density in the region of the carbon nuclei is increased; (2) the electron density along the line connecting a pair of carbon atoms is little changed; (3) there is a large pileup of electrons in the center of the ring; and (4) there is also an increase in electron density just outside of the ring, with this latter effect being most pronounced near the midpoint of each C–C bond. Another plot (not shown) through one of the carbon atoms and the pair of hydrogens bonded to it shows that the C–H bonds are not much affected by the ring closure. A Mulliken population analysis shows that closing the ring (by going from the distorted butane to the cyclopropane) leads to a decrease in C–C overlap population from 0.7 to 0.5 e, as calculated in a (52/2) Gaussian basis set.

J. Hydrogen Sulfide and Its Two Hypothetical Derivatives: H$_2$SO and H$_2$SO$_2$

The molecules sulfur dichloride, SCl$_2$, thionyl chloride, OSCl$_2$, and sulfuryl chloride, O$_2$SCl$_2$, form an interesting series of compounds in which the molecular orbitals of one molecule should be easily correlated with those of another. A similar set of molecules is dimethyl sulfide, (CH$_3$)$_2$S, dimethyl sulfoxide, (CH$_3$)$_2$SO, and dimethyl sulfone, (CH$_3$)$_2$SO$_2$. Because of the relatively large basis set needed to describe the molecules with chlorine atoms or methyl groups, we decided to carry out a study of the related series of molecules in which the pair of chlorine atoms or methyl groups was replaced by hydrogen atoms to give the following series: hydrogen sulfide, H$_2$S, sulfur hydrate, H$_2$SO, and tautomeric sulfoxylic acid, H$_2$SO$_2$. The latter two molecules are not known. These three molecules were studied [11] in a Gaussian basis set with three s-type functions allowed to each hydrogen, five s- and two p-type

functions to each oxygen, and nine s-, five p-, and one d-type function to the sulfur. Molecular optimization of the d exponent was carried out on both H$_2$S and H$_2$SO$_2$, and the average value was used for H$_2$SO.

Electron-density plots were obtained for the filled valence-shell molecular orbitals in the plane delineated by the sulfur and the two hydrogen atoms bonded to it as well as in the perpendicular plane passing through the sulfur and bisecting the distance between the hydrogens. These plots are shown in Figs. 3.34–3.39. The molecule H$_2$S has four filled molecular orbitals in its valence shell, H$_2$SO has seven in its valence shell, and H$_2$SO$_2$ has ten. The energies of these orbitals are compared in Fig. 3.40, in which the predominant contribution to each orbital is also indicated. The series of molecular orbitals, 4a$_1$ and 2b$_2$ for H$_2$S, 7a$'$ and 2a$''$ for H$_2$SO, and 6a$_1$ and 2b$_2$ for H$_2$SO$_2$, are dominated by S–H bonding. Likewise, orbital 6a$'$ for H$_2$SO and orbitals 5a$_1$ and 3b$_2$ for H$_2$SO$_2$ are dominated by S–O bonding. The remaining orbitals exhibit predominantly lone-pair character which, of course, should be entirely sulfur lone pairs for H$_2$S and entirely oxygen lone pairs for H$_2$SO$_2$. In Fig. 3.40, S or O indicates which kind of lone pair is involved.

Now let us view the molecular orbitals of these three molecules in the HSH plane. Note in Figs. 3.34–3.36 that orbitals 4a$_1$ of H$_2$S, 7a$'$ of H$_2$SO, and 6a$_1$ of H$_2$SO$_2$ all show strong S–H bonding, with the sulfur contribution to this bonding being dominated by the 3s atomic orbitals of the sulfur, whereas orbitals 2b$_2$ of H$_2$S, 2a$''$ of H$_2$SO, and 2b$_2$ of H$_2$SO$_2$ exhibit S–H bonding involving sideways interaction of each of a pair of sulfur 3p lobes with each hydrogen. Orbitals 5a$_1$ of H$_2$S and 8a$'$ of H$_2$SO are ascribed to the sulfur lone pair, whereas orbital 7a$_1$ of H$_2$SO$_2$ has been noted in Fig. 3.40 as being dominated by the oxygen lone pair. However, in Figs. 3.31–3.33, it is seen that in the HSH plane these three orbitals are very much alike. Likewise, orbitals 2b$_1$ of H$_2$S, 9a$'$ of H$_2$SO, and 4b$_1$ of H$_2$SO$_2$ are seen to be similar in that they exhibit a nodal plane in the HSH plane. Because of the large number of orbitals associated with H$_2$SO and H$_2$SO$_2$, Figs. 3.35 and 3.36 are continued in the right column of page 59 from which it can be seen that orbital 10a$'$ of H$_2$SO and 8a$_1$ of H$_2$SO$_2$ are similar in the HSH plane and that this is also true of 3a$''$ of H$_2$SO and 3b$_2$ of H$_2$SO$_2$.

In Figs. 3.37–3.39, the orbitals of these three related molecules are shown in the plane perpendicular to that of Figs. 3.34–3.36, a plane

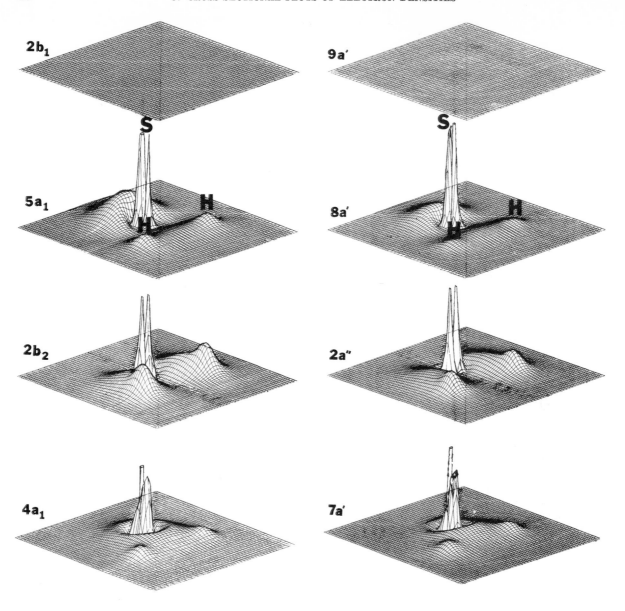

Fig. 3.34. Cross-sectional electron-density plots of the valence-shell molecular orbitals of hydrogen sulfide, H_2S, in a plane passing through the sulfur and the two hydrogen atoms.

Fig. 3.35a and b. Cross-sectional electron-density plots of the valence-shell molecular orbitals of sulfur hydrate, H_2SO, in a plane passing through the sulfur and the two hydrogen atoms. See page 59 for Fig. 3.35b.

Fig. 3.35a.

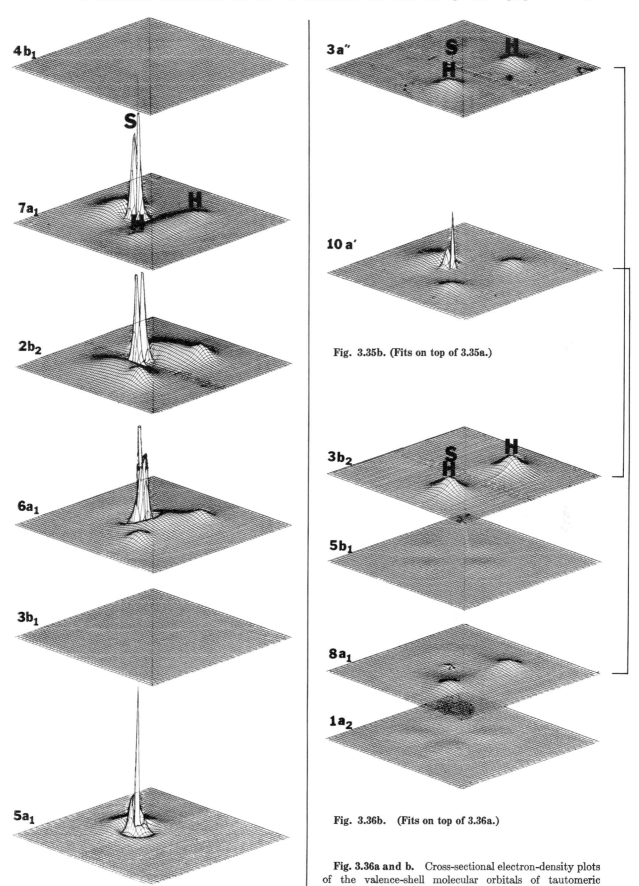

Fig. 3.35b. (Fits on top of 3.35a.)

Fig. 3.36b. (Fits on top of 3.36a.)

Fig. 3.36a.

Fig. 3.36a and b. Cross-sectional electron-density plots of the valence-shell molecular orbitals of tautomeric sulfoxylic acid, H₂SO₂, in a plane passing through the sulfur and the two hydrogen atoms.

that passes through the sulfur and bisects a line connecting the two hydrogen atoms. The orbitals $4a_1$ of H_2S, $7a'$ for H_2SO, and $6a_1$ for H_2SO_2 exhibit no S–O bonding (to be seen in the plane of these diagrams), although orbitals $7a'$ of H_2SO and $6a_1$ of H_2SO_2 do show oxygen lone-pair character. All three of these molecular orbitals are obviously closely related, as shown in Figs. 3.34–3.39. Molecular orbital $5a_1$ of H_2S is dominated by the sulfur lone pair, whereas orbital $8a'$ of H_2SO exhibits both oxygen and sulfur lone-pair character and orbital $7a_1$ of H_2SO_2 is dominated by oxygen lone-pair character. Obviously, these three orbitals belong to the same set and involve the sulfur 3p atomic orbital exhibiting lobes directed along the axis that bisects a line connecting the two hydrogen atoms. Similarly, orbitals $2b_1$ of H_2S, $9a'$ of H_2SO, and $4b_1$ of H_2SO_2 are interrelated, being dominated by sulfur lone-pair character in the case of H_2S and oxygen lone-pair character in the case of H_2SO_2. Note that orbital $7a_1$ of H_2SO_2 still exhibits quite a bit of what may be called sulfur lone-pair character which, in the case of this particular orbital, gives a bridge of charge between the two oxygen atoms. The molecular orbitals participating in S–O bonding will obviously be more stable than those involved in S–H bonding. These S–O dominated orbitals are the related pairs $6a'$ for H_2SO and $5a_1$ for H_2SO_2, as well as $3b_1$ for the latter molecule.

The less stable filled orbitals of H_2SO and H_2SO_2 are plotted in the continuation of Figs. 3.38–3.39. All of these orbitals are dominated by oxygen lone-pair character, as is obvious from the electron-density plots of orbitals $10a'$ of H_2SO and $8a_1$ and $5b_1$ of H_2SO_2. Orbitals $3a''$ of H_2SO and $3b_2$ of H_2SO_2 exhibit no electron density in the chosen plane, which passes through the sulfur and the two oxygen atoms of H_2SO_2. However, the Mulliken population analysis shows that orbital $3a''$ of H_2SO exhibits $1.39\ e$ associated with the oxygen atom (gross population) and a S–O overlap of $0.20\ e$. This means that oxygen 2p electrons are involved in this orbital and that they are so oriented that the plane of Fig. 3.38b is a nodal plane. Not surprisingly, Fig. 3.35b shows detectable electron density only on the hydrogen atoms in the HSH plane. Similarly, orbitals $1a_2$ and $3b_2$ of H_2SO_2 exhibit a gross population of $0.87\ e$ for each oxygen and an electron distribution close to that of orbital $3a''$ of H_2SO, as can be seen from the pertinent electron-density diagrams.

The nodal surfaces appearing in the valence-shell region for the valence molecular orbitals of the H_2S and H_2SO_2 molecules are shown in Fig.

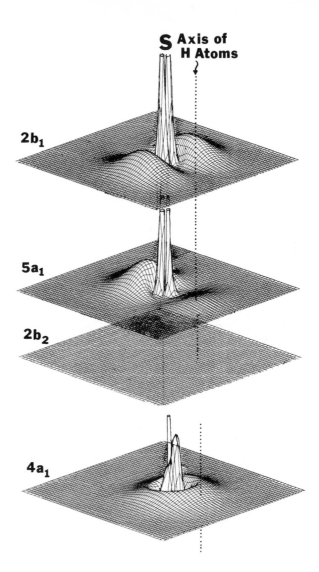

Fig. 3.37. Cross-sectional electron-density plots of the valence-shell molecular orbitals of hydrogen sulfide in a plane passing through the sulfur atom perpendicular to that of Fig. 3.34.

3.41. For both molecules, the most stable valence-shell orbital exhibits no nodal surfaces, except for those s-type, essentially spherical surfaces in the core region that assure orthogonality with the core electrons. For H_2S there are three nodal planes that extend through the valence-shell region. These correspond to the three different 3p orbitals of the sulfur atom. Note that only two of these 3p atomic-orbital nodes lie in symmetry planes of the molecule and hence, change the symmetry designation of the resulting molecular orbital (i.e., an a_1 orbital).

The nodal surfaces passing through the valence region of H_2SO_2 are considerably more complex

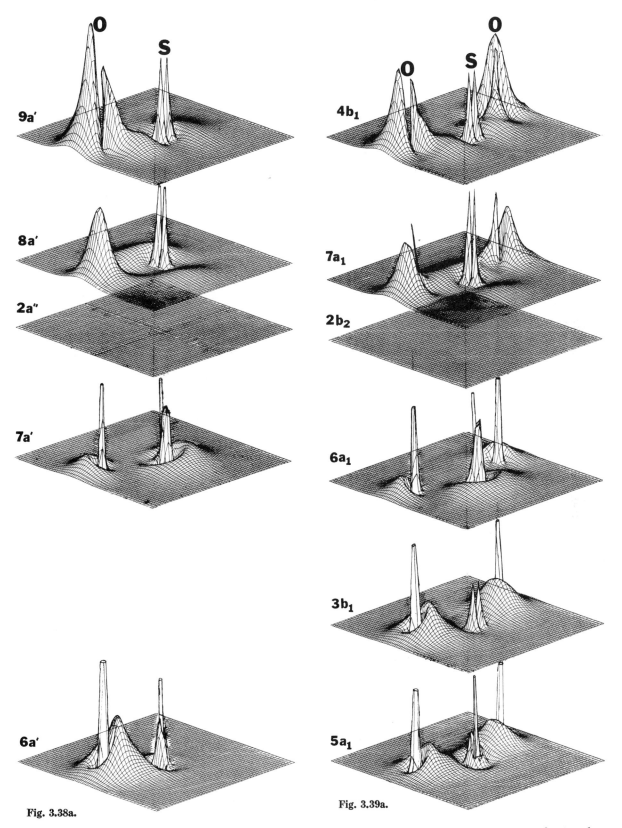

Fig. 3.38a.

Fig. 3.39a.

Fig. 3.38a and b. Cross-sectional electron-density plots of the valence-shell molecular orbitals of sulfur hydrate in a plane passing through the sulfur atom perpendicular to that of Fig. 3.35. See page 62 for Fig. 3.38b.

Fig. 3.39a and b. Cross-section electron-density plots of the valence-shell molecular orbitals of tautomeric sulfoxylic acid for a plane passing through the sulfur atom perpendicular to that of Fig. 3.36. See page 62 for Fig. 3.39b.

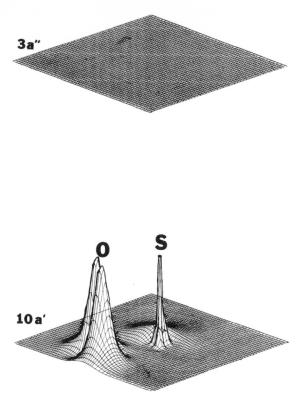

Fig. 3.38b.   (Fits on top of 3.38a.)

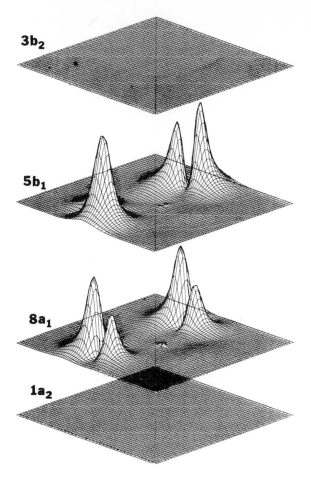

Fig. 3.39b.   (Fits on top of 3.39a.)

than those of H_2S, because the two oxygen atoms utilize their 2p atomic orbitals in the various molecular orbitals of this molecule. Note that molecular orbitals $3b_2$ and $1a_2$ of H_2SO_2 exhibit a pair of mutually perpendicular nodal planes, the intersection of which passes through the sulfur nucleus. This can only correspond to the utilization of the sulfur 3d orbital in the bonding of this molecule. Indeed, the S–O overlap population (0.05 e for orbital $3b_2$ and 0.16 e for $1a_2$) is reduced to zero for orbital $1a_2$ and nearly to zero for orbital $3b_2$ when the calculation is carried out using only s and p atomic orbitals in the basis set. Likewise, the gross electronic population of the sulfur atom for orbital $1a_2$ is similarly reduced to zero and achieves a slight negative value for orbital $3b_2$. In fact, except for orbitals $3b_2$ and $2b_2$, the S–O bonding of all of the molecular orbitals of H_2SO_2 are strongly indebted to contributions from the sulfur 3d atomic orbital; and, for molecular orbitals $3b_2$, which exhibits nodal planes corresponding to d symmetry for the sulfur, the S–H bonding is dominated by the sulfur d character so that the S–H bond is predominantly $(d_\sigma–s_\sigma)$. The reader is invited to lay out the nodal surfaces for the H_2SO molecule, using the electron-density plots of Figs. 3.35 and 3.38.

Fig. 3.40.

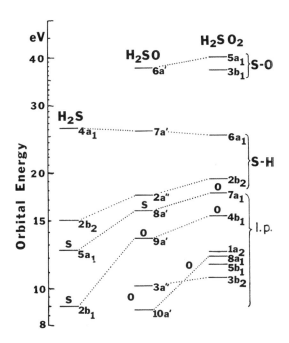

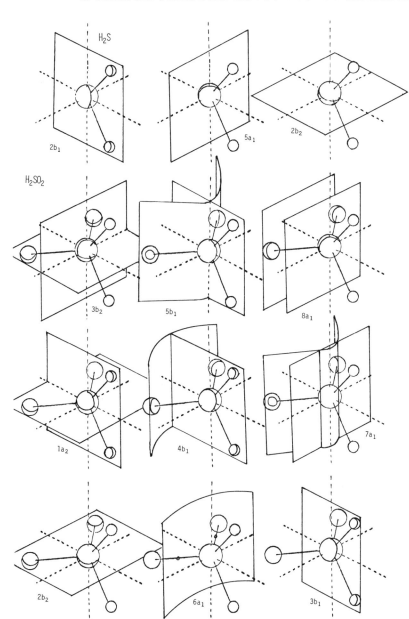

Fig. 3.41. Diagrammatic representation of the nodal surfaces extending into the valence-shell region for the valence orbitals of the H₂S and H₂SO₂ molecules. The central ball stands for the sulfur; the larger outer balls represent oxygen and the smaller ones, hydrogen atoms.

Fig. 3.40. Diagram of the orbital energies of the valence-shell molecular orbitals of H₂S, H₂SO, and H₂SO₂. On the righthand side of this diagram, the predominate contribution to the orbitals is noted (i.e., S—O or S—H bonding, as well as S or O lone-pair character).

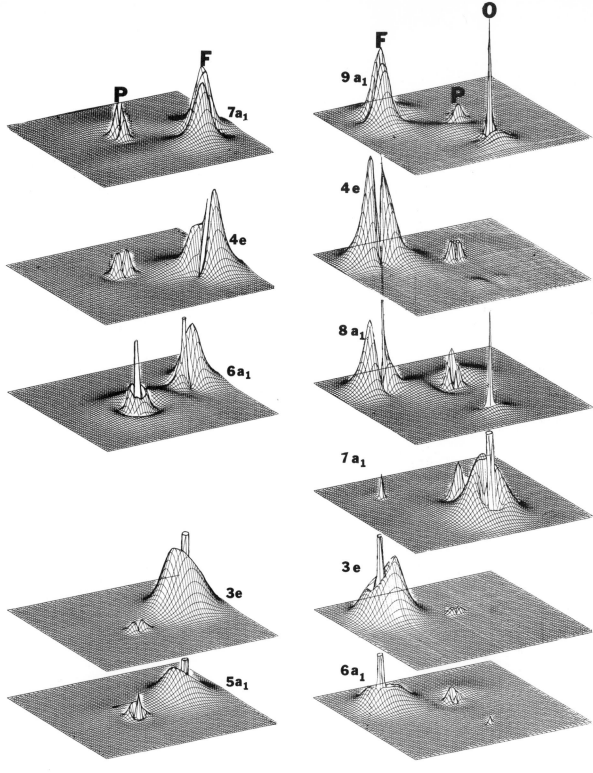

Fig. 3.42a.

Fig. 3.43a.

Fig. 3.42a and b. Cross-sectional electron-density plots for the valence-shell molecular orbitals of trifluorophosphine in a plane passing through the phosphorus, one of the fluorine atoms, and the C_3 axis of the molecule. See page 66 for Fig. 3.42b.

Fig. 3.43a and b. Cross-sectional electron-density plots for the valence-shell molecular orbitals of trifluorophosphine oxide in a plane passing through the phosphorus, the oxygen, one of the fluorine atoms, and the C_3 axis of the molecule. See page 66 for Fig. 3.43b.

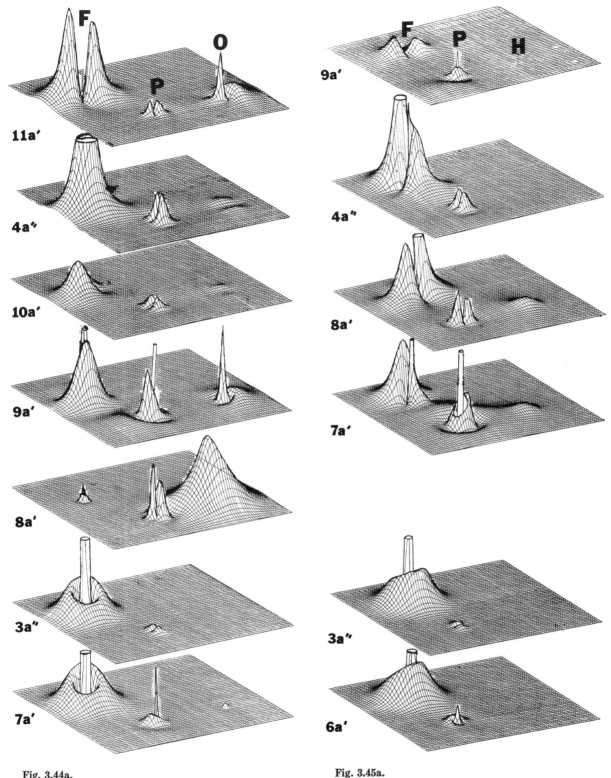

F

O

P

11a′

4a″

10a′

9a′

8a′

3a″

7a′

F

P

H

9a′

4a″

8a′

7a′

3a″

6a′

Fig. 3.44a.

Fig. 3.45a.

Fig. 3.44. Cross-sectional electron-density plots for the valence-shell molecular orbitals of difluorophosphine oxide in a plane passing through the phosphorus, the oxygen, one of the fluorine atoms, and the C₃ axis of the molecule. See page 67 for Fig. 3.44b.

Fig. 3.45a and b. Cross-sectional electron-density plots for the valence-shell molecular orbitals of difluorophosphine in a plane passing through the phosphorus, one of the fluorine atoms, and the hydrogen atom. See page 67 for Fig. 3.45b.

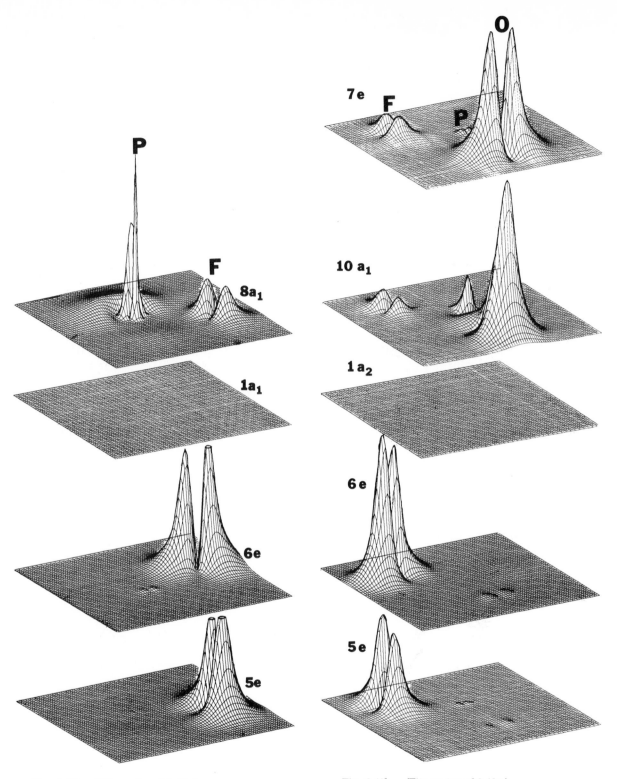

Fig. 3.42b.   (Fits on top of 3.42a.)

Fig. 3.43b.   (Fits on top of 3.43a.)

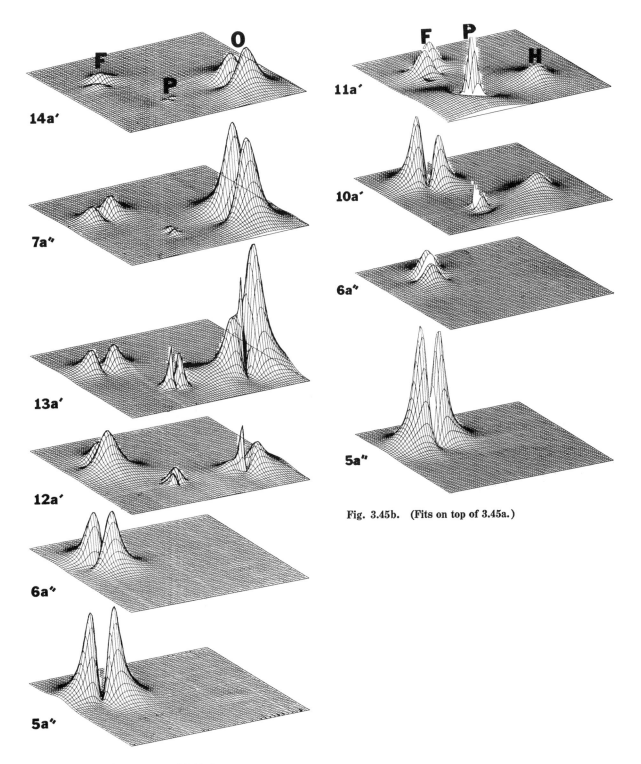

Fig. 3.44b. (Fits on top of 3.44a.)

Fig. 3.45b. (Fits on top of 3.45a.)

67

Table 3.II   Intermolecular Relationships among SCF Delocalized Molecular Orbitals[a]

Major contribution	PF₃	OPF₃	PF₂H	OPF₂H	PH₃	OPH₃
(P–F)σ	5a₁ ↔	6a₁	6a′ ↔	7a′		
(P–F)σ	3e (a₂)	3e (a₂)	3a″ ↔	3a″		
(P–F)σ	3e (a₁)	3e (a₁)				
(P–O)σ		7a₁ ←⟶		8a′ ←⟶		5a₁
(P–H)σ, (P–F)	6a₁	8a₁	7a′	9a′		
(P–H)σ, (P–F)σ			8a′	10a′ ↮	4a₁	6a₁
(P–H)σ					2e (a₂)	2e (a₂)
(P–H)σ					2e (a₁)	2e (a₁)
(P–F)σ, F lp	4e (a₂)	4e (a₂)	4a″ ↔	4a″		
(P–F)σ, F lp	4e (a₁)	4e (a₁)				
(P–F)σ, F lp	7a₁	9a₁	9a′	11a′		
F lp	5e (a₂)	5e (a₂)	5a″	5a″		
F lp	5e (a₁)	5e (a₁)				
F lp	6e (a₂)	6e (a₂)	6a″	6a″		
F lp	6e (a₁)	6e (a₁)				
F lp, (P–H)σ			10a′	12a′ ←/⟶		6a₁
F lp	1a₂ ↔	1a₂				
O lp		10a₁		13a′ ←/⟶		7a₁
P lp, (P–H)σ*		8a₁		11a′		5a₁
O lp, (P–O)π		7e (a₂)		7a″		3e (a₂)
O lp, (P–O)π		7e (a₁)		14a′		3e (a₁)

[a] The double-headed arrows, ⟵⟶, indicate pairs of orbitals that are very closely related, whereas the crossed-off double arrow, ⟵/⟶, shows orbitals that are poorly related.

K. Phosphorus Trifluoride and Phosphoryl Trifluoride, as well as Difluorophosphine and Difluorophosphine Oxide

The molecules PF₃, OPF₃, PF₂H, and OPF₂H have much in common and a number of their molecular orbitals should be readily correlatable as a group and, with even greater similarity, as the closely related pairs of these molecules. In Table 3.II, an interrelation diagram is presented for the valence-shell molecular orbitals of these four molecules. Figures 3.42–3.45, each in two different sections (a and b), show the cross-sectional electron density plots for these molecules, which are displayed in somewhat different orientations. All of these figures are based on calculations [2, 12, 13] involving nine s-, five p-, and one d-type Gaussian functions on the phosphorus, five s- and two p-type on the oxygen and on each fluorine atom, and three s-types on the hydrogen.

From inspection of Figs. 3.42–3.45, using Table 3.II as a guide, the reader is invited to analyze the individual orbitals and to demonstrate the basis for the intermolecular interrelation of the filled valence-shell molecular orbitals. Please note that the cross-sectional plane chosen for the PF₂H molecule passes through the phosphorus, the hydrogen, and one of the fluorine atoms; for the three other molecules, however, it passes through the phosphorus, one of the fluorine atoms, and either the oxygen or (in the case of PF₃) the phosphorus lone pair.

L. Orbital Interrelationships among Water, Formaldehyde, and Ketene

The molecules H_2O, H_2CO, and H_2CCO all exhibit C_{2v} symmetry and are terminated with the same atoms. Indeed, H_2CO and H_2CCO may be regarded as resulting from the progressive introduction of carbon atoms between the H_2 and the O ends of the water molecule. Because of this there should be considerable similarity between their molecular orbitals. In the C_{2v} point group there are four possible symmetry representations (A_1, A_2, B_1, and B_2) but the A_2 will not appear in any of these essentially cylindrical molecules, in which the C_{2v} symmetry is established by the presence of the pair of hydrogen atoms. Note that

Fig. 3.47. Cross-sectional electron-density plots in the molecular plane showing the total valence-shell electrons of water, formaldehyde, and ketene.

Fig. 3.46. Cross-sectional electron-density plots in the molecular plane showing the total electronic structure of water, formaldehyde, and ketene.

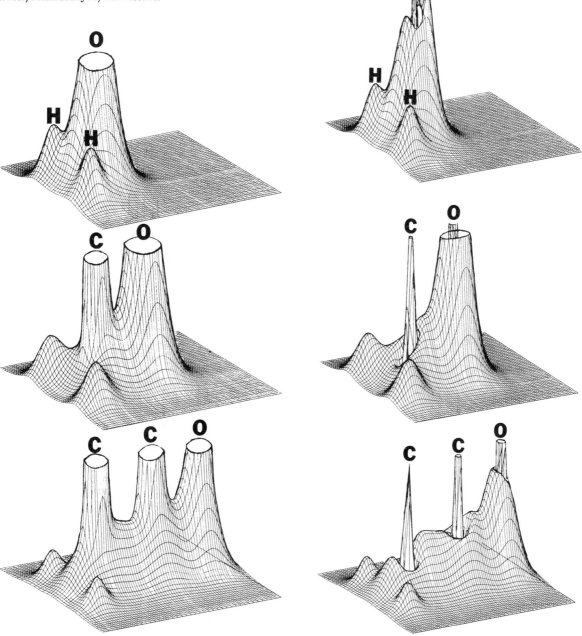

with respect to the C_2 axis, the B_1 and B_2 representations can be interchanged depending on the arbitrary placement of the coordinate axis with respect to the pair of hydrogen atoms. In the following comparisons the water, formaldehyde, and ketene molecules are similarly aligned with their hydrogen atoms in the same orientation. The electron-density plots for these molecules are based on the use [2] of five s- and three p-type atom-optimized Gaussian functions per carbon or oxygen atom and three s-types for each hydrogen atom in the LCAO–MO–SCF computations.

Diagrams of the total and the valence-shell electron densities are shown, respectively, in Figs.

3.46 and 3.47 for water, formaldehyde, and ketene reading from top to bottom. Electron-density diagrams of the molecular orbitals exhibiting A symmetry are presented in Figs. 3.48–3.50, which correspond to water, formaldehyde, and ketene, respectively. The diagrams of these figures represent cross sections through the plane established by the C_2 axis of each molecule and the two

Fig. 3.49. Cross-sectional electron-density plots in the molecular plane of the valence-shell formaldehyde molecular orbitals of A_1 symmetry.

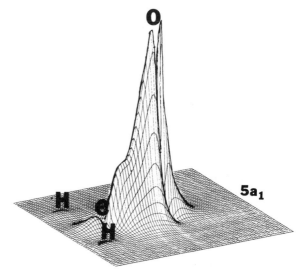

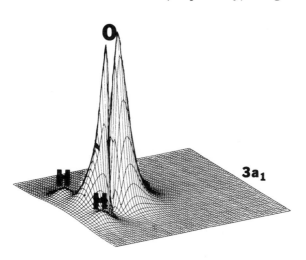

Fig. 3.48. Cross-sectional electron-density plots in the molecular plane of the valence-shell water molecular orbitals of A_1 symmetry.

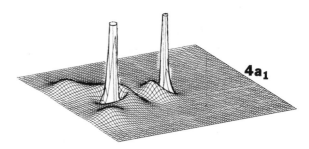

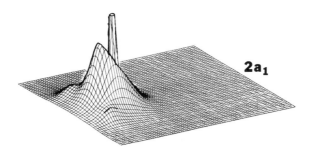

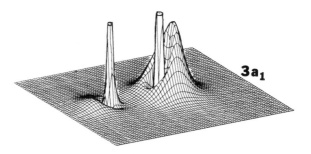

hydrogen atoms. Figures 3.51–3.53, the graphs of which correspond to the B_1 representation, depict the electron densities in a plane that passes through the C_2 axis of each molecule at right angles to the one described above.

The valence-orbital plots of Fig. 3.47 are interesting in that they give an indication of bond polarity. Obviously the oxygen in water is electron-withdrawing with respect to its two neighboring hydrogen atoms. Moreover, in formalde-

Fig. 3.50. Cross-sectional electron-density plots in the molecular plane of the valence-shell ketene molecular orbitals of A_1 symmetry.

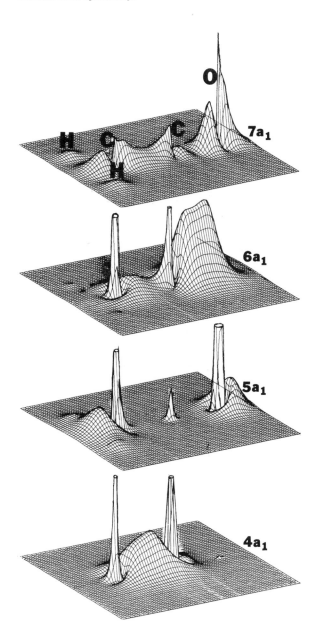

hyde, the oxygen is seen to withdraw electrons from the carbon, which in turn withdraws them from the hydrogen atoms. However, the valence-shell electron-density plot for ketene turns out to be different than might be predicted from the usual notions of chemistry, in that the slope of the charge-density envelope in the region between the atom centers indicates little polarity in the bonds. Indeed, the methylene carbon appears to be slightly electron withdrawing with respect to the carbonyl carbon. Both the total and the valence-orbital plots show a progressive diminution in the intensity of electronic charge in the region of the hydrogen nuclei when going from H_2O to H_2CO to H_2CCO.

Because the plots in Figs. 3.48–3.53 are grouped with respect to symmetry rather than in the usual sequence of increasing stability, the reader is referred to Fig. 3.54 for a graph of the energies of these filled valence-shell molecular orbitals. This figure indicates the orbital mixing involved in interrelating these orbitals, including some admixture of certain virtual orbitals. In order to emphasize the fact that the electron-density plots of the orbitals of B_1 symmetry (shown in Figs. 3.51–3.53) correspond to a cross section through the molecular plane which is perpendicular to that containing the pair of hydrogen atoms of each molecule, the C_2 axes of these B_1 plots are shown as pointing at right angles to their direction in the electron density diagrams for the B_2 orbitals as well as in the diagrams of the orbitals of A symmetry in Figs. 3.48–3.50.

All of the orbitals in the B_1 representation correspond to utilization of the $2p$ lobes of the oxygen (and, when pertinent, of the carbon) that extend perpendicularly to the molecular plane containing the two hydrogen atoms.

In Figs. 3.48–3.50, it can be seen that the $3a_1$ orbital of H_2O, the $5a_1$ of H_2CO, and the $7a_1$ of H_2CCO represent σ interactions (with respect to the C_2 axis of each molecule) involving the $2p$ atomic orbitals of the oxygen and, where pertinent, of the carbon atoms. In contrast, orbital $2a_1$ of H_2O, $3a_1$ and $4a_1$ of H_2CO, and $4a_1$ and $5a_1$ of H_2CCO correspond to σ interactions between the s orbitals on the various atoms making up the molecule. Orbital $6a_1$ of ketene uses both p and s atomic orbitals in its bonding structure, which is also totally σ. Not surprisingly, its energy lies partway between the energies of the $4a_1$ and $5a_1$ orbitals of formaldehyde, as shown in Fig. 3.54. The reason the energy of orbital $5a_1$ of ketene lies partway between those of orbitals $3a_1$ and

$4a_1$ of formaldehyde is apparent from inspection of Figs. 3.49 and 3.50. Likewise, one would expect a gradual stabilization, as shown in Fig. 3.54, when going from orbital $2a_1$ of water to $3a_1$ of formaldehyde to $4a_1$ of ketene, each of which is the valence orbital of lowest energy for its respective molecule. In this sequence, the electronic

charge is stabilized by being spread out (delocalized) over an increasing number of nuclei.

In Figs. 3.51–3.53, it can be seen that the $1b_1$ plot for water obviously corresponds to one of the lone pairs of electrons of the oxygen atom. This molecular orbital transforms to orbital $1b_1$ in formaldehyde, in which this lone pair on the oxygen of the original water molecule is seen to form a π bond with the inserted carbon atom.

Fig. 3.51. Cross-sectional electron-density plots of the valence-shell water molecular orbitals of B symmetry, with the b_2 orbital in the molecular plane and the b_1 orbital in the perpendicular plane.

Fig. 3.52. Cross-sectional electron-density plots of the valence-shell formaldehyde molecular orbitals of B symmetry, with the b_2 orbitals in the molecular plane and the b_1 orbital in the perpendicular plane passing through the carbon and the oxygen atoms.

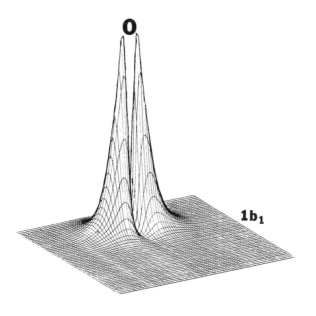

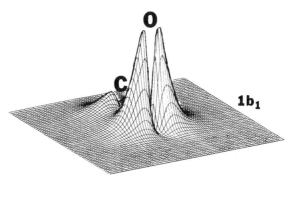

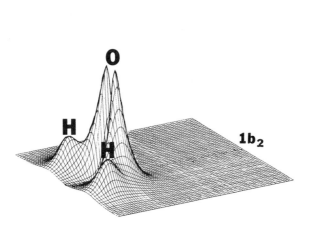

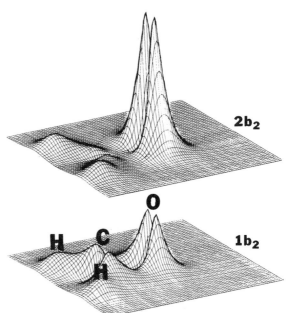

Orbital $1b_1$ in ketene continues this pattern so that (as shown in Fig. 3.54) there is a continuing stabilization of this orbital when going from H_2O to H_2CO to H_2CCO, because of the increased delocalization resulting from stepwise addition of another atom to the molecule. However, molecular orbital $2b_1$ of ketene is destabilized with respect to orbital $1b_1$ of formaldehyde, since orbital $2b_1$ of ketene shows antibonding between the oxygen and its carbonyl carbon. The introduction of the additional node due to antibonding causes localization of charge and this results in destabilization, even though the ketene molecule has one more atom than does formaldehyde. Thus, the $2b_1$ orbital of ketene should be correlated with the first empty orbital, $2b_1$, of formaldehyde. This formaldehyde virtual orbital, $2b_1$, is π^* with respect to the C–O bond, as is orbital $2b_1$ of ketene.

The molecular orbitals of the B_2 representation include the π interactions with respect to the C_2 axis of these molecules, with the nodal plane bisecting the line connecting the two hydrogen atoms, just as the π orbitals subsumed in the B_1 representation correspond to the nodal plane

Fig. 3.53. Cross-sectional electron-density plots of the valence-shell ketene molecular orbitals of B symmetry, with the b_2 orbitals in the molecular plane and the b_1 orbital in the perpendicular plane passing through the two carbon and the oxygen atoms.

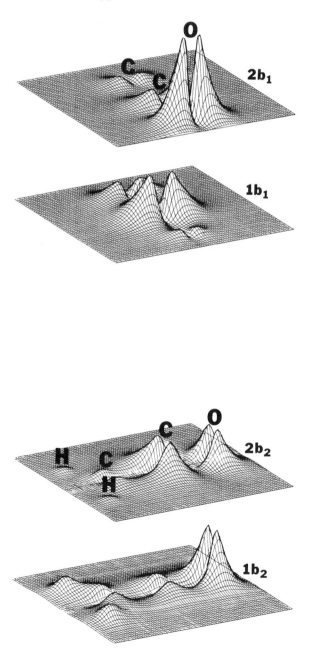

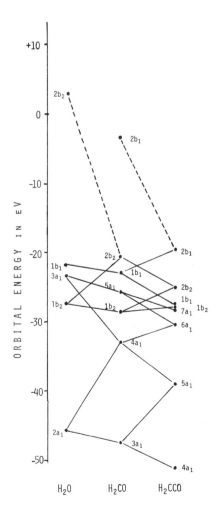

Fig. 3.54. Orbital energies and the interrelationships between the valence-shell molecular orbitals of water, formaldehyde, and ketene.

passing through the two hydrogen atoms. In the water molecule, orbital $1b_2$ corresponds to O–H (p_σ–s_σ) bonding and this bonding character is also found for the C–H bond of orbitals $1b_2$ and $2b_2$ of formaldehyde and $1b_2$ and $2b_2$ of ketene. The $1b_2$ orbitals of H_2O, H_2CO, and H_2CCO all exhibit about the same energy, as can be seen from Fig. 3.54, whereas orbital $2b_2$ of H_2CO is less stable than $1b_2$ of H_2O and $2b_2$ of H_2CCO. This destabilization is effected by the introduction of an additional node to orbital $2b_2$ of formaldehyde caused by its antibonding character for the bond between the carbon and oxygen atoms. Since there is a greater electronegativity difference between oxygen and hydrogen than between carbon and hydrogen, a comparison of the electron density along each O–H bond axis in orbital $1b_2$ of H_2O (see Fig. 3.51) with that along the C–H bond in orbitals $1b_2$ of H_2CO (Fig. 3.52) and of H_2CCO (Fig. 3.53) shows that there is much more stablization caused by O–H overlap in H_2O than by C–H overlap in H_2CO and H_2CCO. However, in both H_2CO and H_2CCO, further stablization is attainable through delocalization by a π-type interaction between the 2p orbitals of the relevant carbons and oxygens.

M. Monomeric Borane and Diborane

At room temperature the equilibrium between monomeric borane, BH_3 and diborane, B_2H_6, is shifted far toward the diborane, with a heat of -40 kcal/mole for the formation of B_2H_6 from 2 moles of BH_3. However, monomeric borane is an interesting molecule because it is planar and has only three filled valence-shell molecular orbitals. A plot of the total electron density of the BH_3 molecule, as measured in the molecular plane, is presented at the top of Fig. 3.55, which shows the electron densities obtained from an *ab initio* calculation [1] based on a minimum-Slater description of the atoms.

The most stable of the three filled valence-shell molecular orbitals, $2a_1'$ of BH_3, is shown as the bottom diagram in Fig. 3.55, from which it can be seen that this orbital involves the interaction of the s valence-shell atomic orbitals of the boron and the three hydrogen atoms. The next two molecular orbitals of boron represent a degenerate pair of E' symmetry. This pair of $1e'$ orbitals represents the interaction with the hydrogen 1s orbitals of the set of boron 2p orbitals lying in the molecular plane. Taken together, the $1e'$ molecular orbitals give a ring of charge around the boron

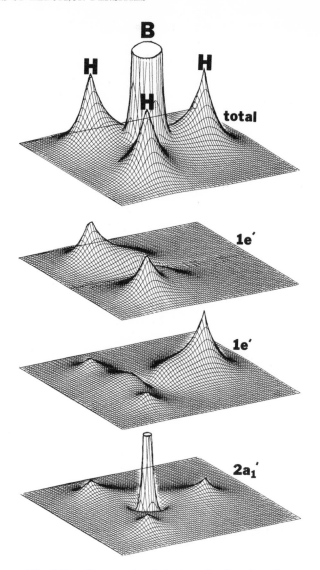

Fig. 3.55. Cross-sectional electron-density plots showing the total electrons and valence-shell orbitals of borane in the molecular plane.

interacting with the three equally spaced hydrogen atoms lying in the same plane.

In diborane [1], each of the two boron atoms is approximately tetrahedrally surrounded by four hydrogen atoms, with two of the hydrogens on each boron being shared with the other boron. This means that there are two planes of interest in B_2H_6. One of them is the plane containing the bridging hydrogens,

$$\begin{array}{ccc} & H & \\ B & & B \\ & H & \end{array}$$

and the other is the plane perpendicular to it.

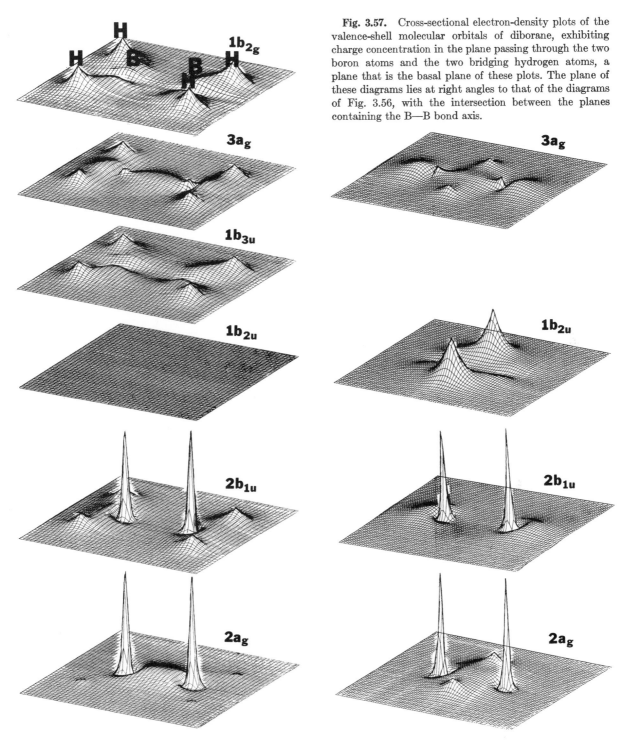

Fig. 3.57. Cross-sectional electron-density plots of the valence-shell molecular orbitals of diborane, exhibiting charge concentration in the plane passing through the two boron atoms and the two bridging hydrogen atoms, a plane that is the basal plane of these plots. The plane of these diagrams lies at right angles to that of the diagrams of Fig. 3.56, with the intersection between the planes containing the B—B bond axis.

Fig. 3.56. Cross-sectional electron-density plots showing the valence-shell molecular orbitals of diborane, B_2H_6, in the plane passing through the two boron atoms and the four terminal hydrogen atoms.

Figure 3.56 depicts the electron density in the perpendicular plane, which contains the two boron atoms and the two pairs of hydrogen atoms lying

at either end of the molecule. Figure 3.57 depicts the plane containing the bridging hydrogen atoms. All but one, the $1b_{2u}$, of the eight filled valence-shell molecular orbitals exhibit electron density in the plane of Fig. 3.56, whereas in the perpendicular plane containing the two bridging hydrogen atoms, electron density is seen for only four of these molecular orbitals, $2a_g$, $2b_{1u}$, $1b_{2u}$, and $3a_g$. Note that the most stable of the valence-shell molecular orbitals of diborane, $2a_g$, involves atomic s-orbital interaction between all eight atoms. As would be expected, a considerable proportion of the electron density of this orbital is found within the tetratomic ring formed by the boron and bridging hydrogen atoms.

Orbital $2b_{1u}$ also involves the s orbitals but is antibonding in the region between the two boron atoms so that there is no electron density on the bridging hydrogens. The main function of this orbital is to bond each boron to its pair of nonbridging hydrogens. The next less stable orbital, $1b_{2u}$, represents π bonding between the two boron atoms and, of course, these boron p orbitals interact strongly with the bridging hydrogens so as to give good bonding between them and the boron atoms. Orbital $1b_{3u}$ corresponds to some B–B π-like interaction in the plane that is perpendicular to the two bridging hydrogen atoms, with the main function of this orbital being the contribution of strong bonding interactions between each boron atom and its pair of nonbridging hydrogen atoms. Orbital $3a_g$ corresponds to the end-on interaction of the boron p orbitals along the B–B bond axis. As would be expected, this leads to interaction with all six of the hydrogen atoms, with electron density in both planes of interest. The least stable of the filled valence-shell orbitals of B_2H_6 is molecular orbital $1b_{2g}$. This orbital is similar to orbital $1b_{3u}$ except that it has an additional nodal plane, a plane that is perpendicular to and bisects the B–B axis.

N. Chlorosilane and the Role of d Orbitals in Charge Transfer to Third-Period Atoms

The chlorosilane molecule, H_3SiCl, was studied [14] in four different basis sets corresponding to no d character being allowed to either silicon or chlorine, d character on the silicon, d character on the chlorine, and d character on both of these atoms. The corresponding uncontracted Gaussian basis sets were the following, with the listing being in the order of H/Si/Cl: (3/95/95), (3/951/95), (3/95/951), and (3/951/951). The resulting electron-density plots for each of the individual orbitals and for the total molecule differed only slightly from one basis set to another so that, to the casual observer, allowing d character made an inconsequential difference in the gross electronic distributions. However, the addition of d

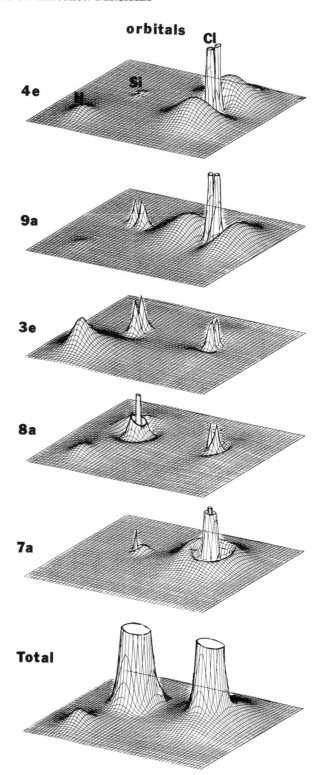

Fig. 3.58. Cross-sectional electron-density plots of the valence-shell orbitals and total electronic distribution for chlorosilane in the plane passing through the chlorine, the silicon, and one of the hydrogen atoms.

Si; d – no d   **Cl; d – no d**

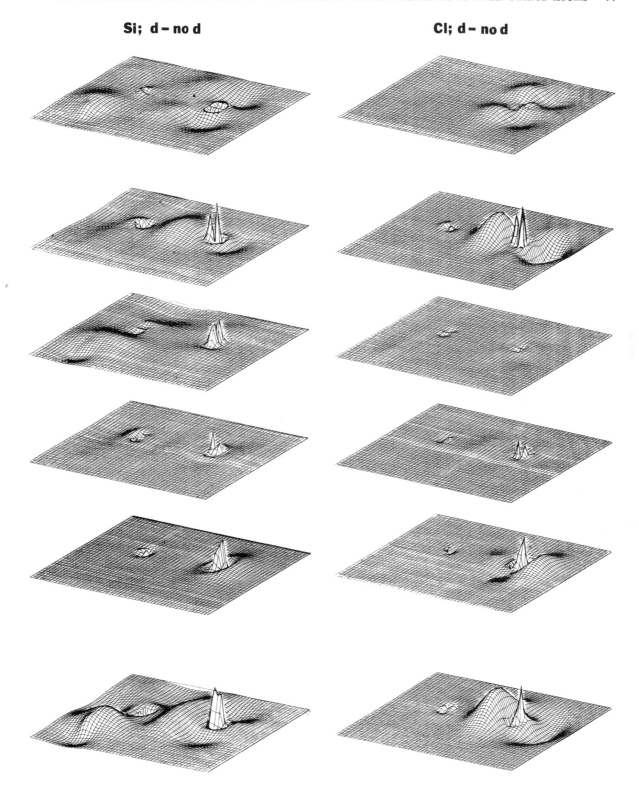

Fig. 3.59. Cross-sectional electron-density difference plots for the chlorosilane molecule corresponding to the electron density calculated with d character on the silicon minus the density calculation with no d character. The plots in this figure correspond to the neighboring plots in Fig. 3.58, except that the electron-density scale is five times more sensitive.

Fig. 3.60. Cross-sectional electron-density difference plots for the chlorosilane molecule corresponding to the electron density calculated with d character on the chlorine minus the density calculation with no d character. The plots in this figure correspond to the plots in Fig. 3.58, except that the electron-density scale is five times more sensitive.

functions led to a considerable change (see Table 3.IV) in such experimental properties as the energy involved in forming the molecule from the separate atoms (i.e., the binding energy) and the dipole moment. Also, there was an appreciable change in the atomic charges and overlap populations obtained from a Mulliken population analysis.

Electron-density plots for the valence-shell orbitals of chlorosilane are presented in Fig. 3.58 along with a plot of the total electron density. These plots correspond to the plane passing through the Si–Cl bond and one of the hydrogen atoms. As expected, the most stable of the valence-shell molecular orbitals, 7a, corresponds to Si–Cl (s_σ–s_σ) bonding and, as most of the charge resides on the chlorine atom, to chlorine lone-pair character. The next orbital, 8a, also involves the s valence orbitals of all of the atoms in the molecule (with appreciable p_σ character on the chlorine), so that the dominant function of this orbital is Si–H (s_σ–s_σ) bonding, with some Si–Cl σ antibonding.

The next pair of orbitals, the 3e set, exhibits π symmetry with respect to the C_3 axis of the molecule (i.e., the Si–Cl bond axis). Again this orbital is dominated by bonding between the silicon and the hydrogen, i.e., Si–H (p_σ–s_σ), but there is also some Si–Cl (p_π–p_π) bonding. The next higher molecular orbital, 9a, corresponds to the use by both the silicon and the chlorine of their respective 3p orbitals having lobes directed along the Si–Cl bond axis. This, of course, leads to Si–Cl (p_σ–p_σ) bonding as well as to appropriate chlorine lone-pair character. As indicated by Fig. 3.58, there is a small charge concentration in the vicinity of the hydrogen atoms that leads to a trace of Si–H bonding in this molecular orbital. The outermost filled molecular orbitals of chlorosilane are the 4e pair, which are chiefly occupied with contributing chlorine lone-pair electrons that have π symmetry with respect to the Si–Cl bond. Because of all of the valence orbitals the pair of 4e orbitals corresponds to the smallest charge

Table 3.III   Orbital Energies and Mulliken Populations for the Valence Shell of H_3SiCl

Orbital[a]	Predominant character[b]	Energy[c] (au)	Gross atomic populations			Overlap population	
			Si	H	Cl	Si–H	Si–Cl
4e[d]	Cl $n_\perp$ + (Si–Cl) π^*	−0.3865	0.0098	0.2081	3.3656	0.0088	−0.0126
Δ_{Si}[e]		−0.0120[f]	+0.3270	−0.0385	−0.2114	+0.0893	+0.0156
Δ_{Cl}		+0.0054	+0.0012	−0.0050	+0.0140	+0.0016	−0.0028
$\Delta_{Si,Cl}$		−0.0071	+0.3140	−0.0418	−0.1884	+0.0867	+0.1464
9a	(Si–Cl) σ + Cl $n_\|$	−0.4393	0.5400	0.0636	1.2692	0.0172	0.2550
Δ_{Si}		−0.0089	+0.0589	−0.0086	−0.0331	+0.0181	+0.0115
Δ_{Cl}		−0.0045	−0.0174	+0.0022	+0.0108	+0.0065	+0.0357
$\Delta_{Si,Cl}$		−0.0091	+0.0352	−0.0036	−0.0244	+0.0208	+0.0460
3e	(Si–H) σ + (Si–Cl) π	−0.5156	1.7312	0.5937	0.4876	0.4521	0.1710
Δ_{Si}		+0.0088	+0.2084	−0.0974	+0.0838	+0.0159	+0.0072
Δ_{Cl}		+0.0082	−0.0464	+0.0172	−0.0052	+0.0001	+0.0012
$\Delta_{Si,Cl}$		+0.0143	+0.1912	−0.0893	+0.0768	+0.0193	+0.0072
8a	(Si–H) σ + (Si–Cl) σ^*	−0.7543	1.1861	0.1934	0.2338	0.2079	−0.0726
Δ_{Si}		+0.0297	+0.0063	−0.0020	−0.0003	+0.0002	+0.0540
Δ_{Cl}		+0.0062	−0.0149	+0.0009	+0.0120	−0.0007	+0.0302
$\Delta_{Si,Cl}$		+0.0327	−0.0036	−0.0060	+0.0216	−0.0031	+0.0817
7a	(Si–Cl) σ + Cl $n_\|$	−1.0865	0.2257	0.0021	1.7681	0.0026	0.2695
Δ_{Si}		−0.0015	+0.0033	+0.0008	−0.0057	+0.0001	+0.0311
Δ_{Cl}		+0.0133	−0.0222	+0.0001	+0.0218	+0.0001	−0.0234
$\Delta_{Si,Cl}$		+0.0078	−0.0209	+0.0004	+0.0196	−0.0001	+0.0008

[a] Note that the values listed for an E symmetry represent the sum of a degenerate pair, i.e., they correspond to an occupancy of four electrons.

[b] The major contribution(s) to the delocalized orbital, with $n_\perp$ and $n_\|$ referring to lone-pair electronic charge, respectively, oriented perpendicular to or along the Si–Cl bond axis.

[c] 1 au = 27.211 eV.

[d] The first line for each orbital presents the values calculated for the (3/95/95) basis set.

[e] Δ_{Si} stands for the (3/951/95) minus the (3/95/95) values, Δ_{Cl} for the (3/95/951) minus the (3/95/95) values, and $\Delta_{Si,Cl}$ for the (3/951/951) minus the (3/95/95) values.

[f] All differences are shown in italics.

concentration at the silicon atom, there is only a small amount of Si–H and Si–Cl interaction, with the latter being π antibonding.

The orbital energies and Mulliken populations for the valence-shell molecular orbitals of chlorosilane are shown in Table 3.III for the calculation not involving d orbitals. Below each of these values are shown in italics the respective changes resulting from allowing d character to the silicon, Δ_{Si} ; to the chlorine, Δ_{Cl} ; and to both the silicon and chlorine, $\Delta_{Si,Cl}$. These changes for the Mulliken populations of the valence-shell orbitals should be compared with the corresponding changes in electronic distributions shown for the Δ_{Si} and Δ_{Cl} cases in Figs. 3.59 and 3.60, respectively. (Note that the vertical scale indicating electron density has been magnified five times for Figs. 3.59 and 3.60 as compared to Fig. 3.58.) Table 3.III shows that allowing d character to the silicon leads consistently to an increase in the gross atomic population of silicon for each of the orbitals, with this transfer of charge to the silicon usually coming from the chlorine atom (except for the pair of 3e orbitals, which also exhibit an increase in charge on the chlorine, and the 8a orbital, in which most of the charge transferred to the silicon comes from the hydrogens). The transfers of charge and changes in overlap popula-

tion shown in Table 3.III for the situation corresponding to allowing d character to the silicon are clearly reflected in Fig. 3.59.

Table 3.III also shows that, when d character is allowed to the chlorine, the transfer of charge is from the silicon to the chlorine in the case of orbitals 7a and 8a, whereas for the remaining valence-shell orbitals the charge transfer is much less than for the situation in which d character is allowed to the silicon and not to the chlorine. Again, the changes given in Table 3.III for atomic and overlap populations for each orbital when d character is allowed to the chlorine are readily related to the corresponding shifts in electron density depicted in Fig. 3.60.

According to the Mulliken population analysis, it is clear from Table 3.IV that, when d character is allowed only to the silicon, there is an overall charge feedback from both the chlorine and hydrogen atoms to the silicon. This feedback amounts to 0.17 e from the chlorine and 0.15 e from each hydrogen, and it occurs concurrently with an increase in both the Si–Cl and Si–H overlap populations. When the silicon d orbitals are considered alone, the Mulliken population analysis shows that 0.54 e is accepted by the d atomic orbitals of π symmetry and 0.07 e by the ones of σ symmetry with respect to the Si–Cl bond. The

Table 3.IV Some Overall Properties Calculated for Chlorosilane

Property	(3/95/95) Basis set plus			
	No d's	d on Si	d on Cl	d on Si and Cl
Total energy (au)	−749.4829	−749.5767	−749.5032	−749.5888
Potential energy (au)				
Nuclear	+86.5223	+86.5223	+86.5223	+86.5223
One-electron	−1957.9463	−1958.2432	−1958.2700	−1958.5064
Two-electron	+372.3994	+372.7790	+372.7213	+373.0262
Kinetic energy (au)	+749.5417	+749.3652	+749.5231	+749.3691
Virial ratio	−1.99992	−2.00028	−1.99997	−2.00029
Binding energy[a] (eV)	9.00	11.56	9.56	11.89
Dipole moment[b] (D)	−0.889	−0.366	−0.246	+0.115
Charge[c] on H (e)	−0.061	+0.085	−0.077	+0.080
Charge on Si (e)	+0.308	−0.298	+0.408	−0.211
Charge on Cl (e)	−0.124	+0.043	−0.177	−0.029
Si–H overlap pop. (e)	0.689	0.810	0.696	0.811
Si–Cl overlap pop. (e)	0.610	0.863	0.651	0.890

[a] The binding energy is the difference between the sum of the total energies of the constituent atoms and the total molecular energy, with 1 au = 27.211 eV and no correction for molecular extracorrelation energy. The experimental binding energy (which includes correlation) is 14.0 eV.

[b] The experimental dipole moment is 1.28, according to L. O. Brockway and I. E. Coop, *Trans. Faraday Soc.* **34**, 1429 (1938). A negative sign for the calculated dipole moment means that the negative end of the molecule is at the chlorine atom (1 au = 2.542 D).

[c] The charge on an atom is obtained by subtracting the Mulliken gross population of that atom from its atomic number.

concurrent increase in the Si–Cl overlap population is 0.16 e for the π system and 0.12 e for the σ.

When d character is allowed only to the chlorine, Table 3.IV indicates that the charge transfer is considerably smaller, consisting of an overall loss of 0.10 e from the silicon, with half of this going to the chlorine and half to the three hydrogens. The corresponding change in overlap populations is also small. When the chlorine d orbitals are considered alone, it is found that the chlorine accepts charge more effectively in the σ manifold, 0.05 e, than in the π manifold of d orbitals, for which the charge change is 0.01 e. As shown in Tables 3.III and 3.IV the changes resulting from allowing d character to both the silicon and chlorine are approximately additive.

The Mulliken population analysis clearly demonstrates that allowing d character to the silicon results in electronic feedback to this atom from both the chlorine and hydrogen atoms, with this effect being dominated by $p_\pi \rightarrow d_\pi$ charge transfer from the chlorine to the silicon. In contrast the transfer of charge is considerably less when d character is allowed to the chlorine. In this case, the Mulliken population analysis indicates that the major effect is one of polarization. However, upon comparing Figs. 3.59 and 3.60, we clearly see that the detailed charge transfer in either of these cases is similar except for the not unexpected fact that, when the d character is given to the silicon, the electron-density plots indicate considerable charge shifts involving the hydrogen atoms, as compared to essentially no such charge shifts when the d character is granted only to the more distant chlorine atom. Note in Figs. 3.59 and 3.60 that for each orbital the effect in the Si–Cl bonding region, as well as in the chlorine lone-pair region, is very similar for a given orbital whether or not the d character is assigned only to the silicon or only to the chlorine. These figures show that the major charge redistribution between the silicon and chlorine is to be found in the following molecular orbitals: 7a, which exhibits an increase in chlorine lone-pair character in both cases; 9a, which shows a shift of σ charge from the chlorine lone-pair to the Si–Cl bonding region; and the 4e pair, which shows a similar shift of π charge.

The fact that the charge shifts upon allowing d character to the silicon may be classified as being dominated by p_π–d_π feedback and upon allowing d character to the chlorine to polarization of this atom, when indeed the two effects are seen to give very similar charge rearrangements, indi-

cates the artificiality of the interpretation usually employed by chemists—an approach in which bonding is discussed in terms of the combination of atomic orbitals used to derive the molecular orbitals in the LCAO approximation. The artificiality of this approximation was pointed out in Chapter 1, so it should not be surprising to find that alternative expansions of the basis set, with respect both to the number of functions and to the angular freedom available to these functions, leads to similar results even though the improvement in the functional description of the molecule has been made in different ways. Thus, the net effect of adding d character to either the silicon or the chlorine is to move charge from the lone-pair region of the chlorine atom to the Si–Cl bonding region, thereby reducing the polarity of the molecule, as indicated by the calculated dipole moments (see Table 3.IV).

O. Phosphorus Pentafluoride

According to the standard chemical-bond picture used by chemists, it is necessary to employ d atomic orbitals (the sp^3d hybrid) to describe pentacoordinate phosphorus. However, it has been known for a number of years that such structures may be equally well described using only s and p atomic orbitals. Two recent sets [15, 16] of $ab\ initio$ LCAO–MO–SCF calculations (both with and without inclusion of d character) on phosphorus pentafluoride, PF_5, have demonstrated that the d atomic orbitals of the phosphorus are not involved in the σ structure of the molecule, although they do contribute considerably to charge feedback from the fluorines to the phosphorus. The largest of these calculations was carried out with a (1061/84) basis set of Gaussian functions contracted to (641/42), with the single d exponent being molecularly optimized. In this study, the geometry was also optimized; and, for the stable trigonal bipyramidal conformation of PF_5, the equatorial and axial P–F bond lengths were found equal to 1.542 and 1.571 Å, respectively, values which are in good agreement with the experimental ones of 1.534 and 1.577 Å. The corresponding no-d calculation was, of course, carried out with a (106/84) contracted to a (64/42) basis set.

In the other LCAO–MO–SCF set of calculations, the emphasis was on inquiring whether or not d character would be involved in the σ structure of the PF_5 molecule if a proportionately large number of d functions were allotted to the phosphorus. In this case, (952/52) and (95/52) basis sets were employed, using the experimental geometry for the trigonal-bipyramidal molecule. The electron-density plots shown in Figs. 3.65–3.72 correspond to this latter set of calculations. Even though ten 3d Gaussian functions were added to a set of 70 sp functions by going from the (95/52) to the (952/52) basis set, there were no gross changes in the electron distribution of any of the molecular orbitals of phosphorus pentafluoride if d atomic orbitals are allowed. Indeed, these changes were all about as subtle as those discussed in the section dealing with chlorosilane. Although the entire contribution from the phosphorus comes from its d orbitals for some of the valence-shell molecular orbitals of phosphorus pentafluoride (e.g., the e″ orbitals), the part of the electron density resulting from this contribution is sufficiently small and diffuse so that the general character of the orbital is not affected by going from the no-d to the with-d situation.

The total energies, as well as the results from Mulliken population analyses, are presented in Table 3.V for the two sets of calculations on the PF$_5$ molecule in its stable geometry. Note that the total energy corresponds to considerably more stability in the case of the calculations employing the large basis sets because the use of only two p-type Gaussian exponents per fluorine atom in

the (52) fluorine basis set is a rather inadequate representation. The addition of two orbital exponents to the (95/52) basis set to incorporate phosphorus d character led to an appreciable increase in the total number of orbitals employed in the basis set, so that the resulting stabilization due to addition of d character was greater than that for the larger basis set, as expected. Similarly, it should be noted that the Mulliken charge on the phosphorus atom was considerably more reduced by addition of the two d exponents to the (95/52) basis set than by inclusion of a single d exponent in the contraction of the (106/84) basis set. Naturally this difference also shows up in the changes in overlap population due to incorporation of d functions. It is clear from the discussion in Chapter 2 that all of these calculations ought to give electron-density plots for the various molecular orbitals (as well as for the total molecule or the sum of the valence orbitals) that differ only in fine details.

Electron-density plots for the total PF$_5$ molecule and its valence shell are shown in Fig. 3.61 for the equatorial plane (top two diagrams) containing the phosphorus and the three equatorial fluorine atoms, as well as for an axial plane passing through the phosphorus, one of these equatorial, and the two axial fluorine atoms (bottom two diagrams). In the companion set of electron-

Table 3.V   Overall Properties Calculated for Phosphorus Pentafluoride

| | | Value from the given Gaussian basis set | | | |
| | | Contraction of | | Uncontracted | |
Property		(106/84)	(1061/84)	(95/52)	(952/52)
Total energy (au)		−837.63	−837.84	−832.50	−832.94
Charge on P (e)		+2.67	+2.12	+2.25	+1.45
Charge on axial F (e)		−0.57	−0.49	−0.46	−0.31
Charge on equat. F (e)		−0.51	−0.40	−0.44	−0.28
P–F$_{ax}$ overlap (e)		0.37	0.56	0.27	0.49
P–F$_{eq}$ overlap (e)		0.39	0.61	0.23	0.47
Atomic orbital occupations					
P	3s		0.74	0.95	0.84
	3p		1.46	1.80	1.58
	3d		0.64	0.00	1.13
F$_{ax}$	2s		1.90	1.99	2.03
	2p$_\sigma$		1.62	1.59	1.60
	2p$_\pi$		3.88	3.88	3.68
F$_{eq}$	2s		1.94	2.00	2.04
	2p$_\sigma$		1.60	1.61	1.59
	2p$_\pi$		3.83	3.83	3.65

density plots, Fig. 3.62, the effect of allowing d orbitals to the molecule is graphically illustrated by a d–(no-d) difference plot shown at a fivefold

magnification of the vertical scale as compared to Fig. 3.61. It is clear from Fig. 3.62 that allowing d character to the phosphorus results in a shift

Fig. 3.61. Cross-sectional electron-density plots for the total and valence-shell electrons of phosphorus pentafluoride, as determined in the equatorial and in an axial plane.

Fig. 3.62. Cross-sectional electron-density difference plots showing the effect of adding d character (d–no d) to the phosphorus atom for the total and valence-shell electrons of phosphorus pentafluoride. The plots in this figure correspond to those in Fig. 3.62, except that the electron-density scale is five times more sensitive.

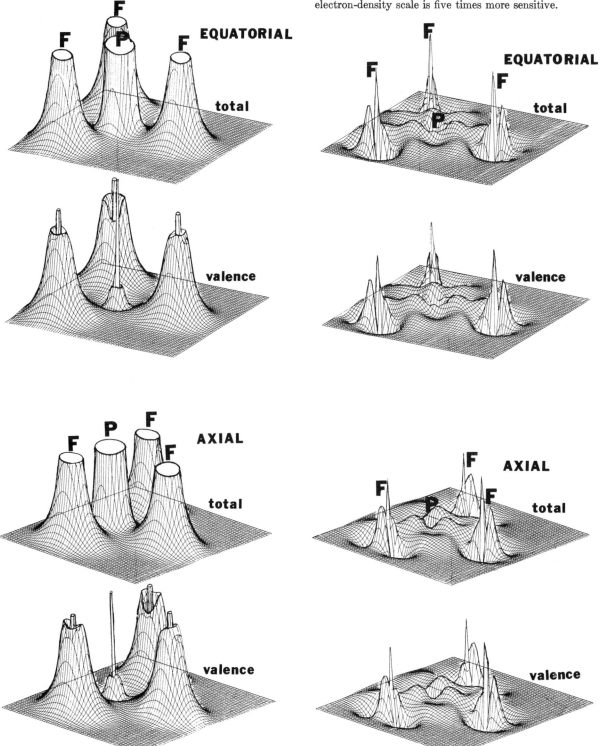

of electronic charge from an annular region around each fluorine atom (with the annulus having the P–F bond as its axis) primarily to the P–F bonding region but also to an approximately spherical region surrounding the phosphorus with charge concentration along the five bonds. These charge shifts are seen to be associated with a diminution

in the amount of charge available to the inner antinodes of the valence orbitals of the phosphorus and, to a lesser extent, of the fluorine atoms.

The valence-shell molecular orbitals of phosphorus pentafluoride lie in two categories with respect to their stability. There is a more stable group of five orbitals ($5a_1'$, the pair of $3e'$, $3a_2''$,

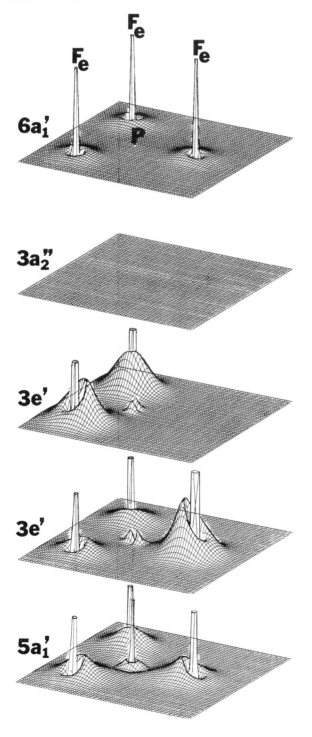

Fig. 3.63. Cross-sectional electron-density plots for the five more stable orbitals of phosphorus pentafluoride, as seen in the equatorial plane.

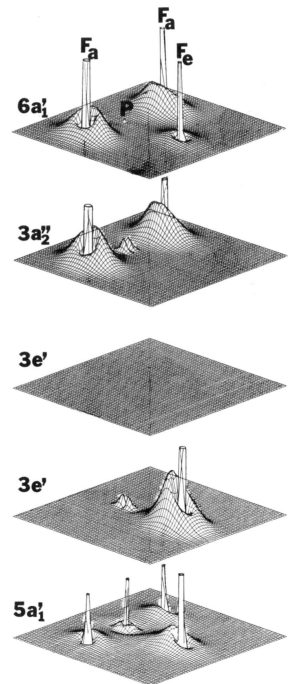

Fig. 3.64. Cross-sectional electron-density plots for the five more stable orbitals of phosphorus pentafluoride, as seen in an axial plane passing through one of the equatorial atoms.

and 6a₁′) and this is followed by a second group consisting of the remaining 15 valence-shell orbitals (7a₁′, 4e′, 4a₂″, 1e″, 5e′, 8a₁′, 1a₂′, 5a₂″, 6e′, and 2e″). Electron-density plots for the more stable group of five orbitals are shown in Figs. 3.63 and 3.64 and for the remaining orbitals in Figs. 3.66–3.69. As can be seen from these figures, the group of the five more stable valence-shell molecular orbitals is differentiated from the remaining valence orbitals by the fact that the former are the only ones dominated by the 2s atomic orbitals of the fluorine atoms.

As can be seen from Fig. 3.63, which represents the equatorial plane of the trigonal-bipyramidal PF₅ molecule, and from Fig. 3.64, which represents an axial plane, orbital 5a₁′ involves the overlap of the s valence-shell atomic orbitals of all of the atoms in the molecule [i.e., F–P (s_σ–s_σ) bonding]. Hence this molecular orbital shows no valence-region nodal surfaces, but only the essentially spherical nodal surfaces of the type exhibited by the fluorine 2s and phosphorus 3s atomic orbitals in the core region. On the other hand, each of the two 3e′ molecular orbitals has a nodal plane in the valence region (passing through the C₃ molecular axis) in addition to the nodal surfaces in the core region. The 3e′ pair of orbitals taken together corresponds to the interaction of the equatorial fluorine 2s orbitals with a ring of charge surrounding the phosphorus nucleus in the equatorial plane of the molecule—a ring of charge due to the appropriate pair of phosphorus 3p atomic orbitals so that there is F_eq–P (s_σ–p_σ) bonding.

Orbital 3a₂″ also has a single nodal plane in the valence region and this molecular orbital corresponds to σ interaction between the axial fluorine 2s atomic orbitals and the phosphorus 3p orbital directed along the C₃ axis of the molecule. The fifth molecular orbital in order of decreasing stability, 6a₁′, exhibits very little charge on the phosphorus atom, with considerable charge on all five of the fluorine atoms. In the (952/52) basis set, orbital 6a₁′ shows P–F antibonding as well as negative overlap between any axial and any equatorial fluorine atom. The presence of a trivially small positive overlap between the three equatorial fluorine atoms and also between the axial fluorines indicates that the valence-region nodal surface for this orbital is shaped like a C₃ᵥ hourglass. Note that orbital 6a₁′ is the only F–P antibonding molecular orbital in the group of the five more stable orbitals of the PF₅ molecule. Furthermore, in the bonding region its nodal surface has roughly the shape of the d atomic orbital which would have been utilized if the σ structure

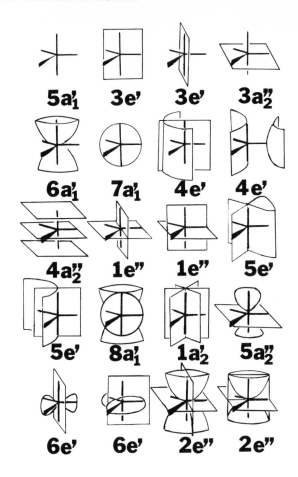

Fig. 3.65. Nodal surfaces appearing in the valence-shell region of the valence-shell molecular orbitals of phosphorus pentafluoride.

of PF₅ were based on an sp³d phosphorus hybrid.

The nodal surfaces in the valence region (excluding the valence-orbital nodal surfaces in the core region that exist to keep the valence orbitals orthogonal to the core orbitals) are shown for the 20 valence-shell orbitals of phosphorus pentafluoride in Fig. 3.65. As previously indicated, orbital 5a₁′ has no valence-region nodes, whereas the pair of 3e′ orbitals plus the 3a₂″ orbital correspond to the three mutually perpendicular nodal planes which may be passed through the central phosphorus atom.

As shown in Figs. 3.66 and 3.67, the next less stable molecular orbital, 7a₁′, involves the phosphorus 3s atomic orbital and 2p orbitals of all of

Fig. 3.66. Cross-sectional electron-density plots for the valence-shell orbitals of intermediate stability of phosphorus pentafluoride, as measured in the equatorial plane.

Fig. 3.67. Cross-sectional electron-density plots for the valence-shell orbitals of intermediate stability of phosphorus pentafluoride, as measured in the axial plane.

Fig. 3.66.

Fig. 3.67.

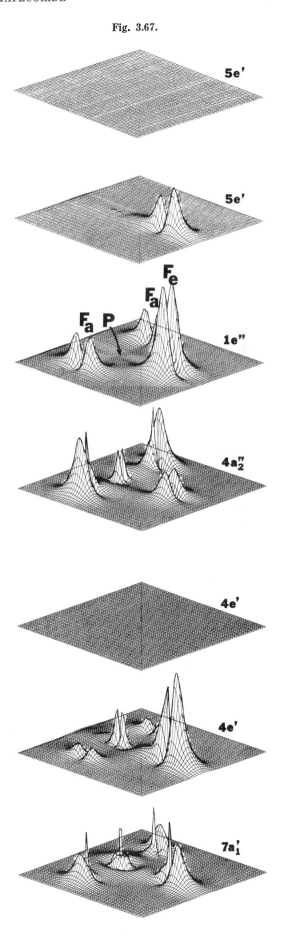

the fluorine atoms, with these 2p lobes being consistently directed toward the phosphorus atom, so that each F–P bond may be described as $(p_\sigma\text{–}s_\sigma)$. Orbital $7a_1'$ exhibits a single essentially spherical outer-nodal surface in the valence region, a nodal surface which passes through each of the five fluorine nuclei.

As shown in Fig. 3.65, all of the remaining orbitals of the PF_5 molecule exhibit two or three valence-region nodal surfaces. The pair of $4e'$ orbitals taken together contributes F–P $(p_\sigma\text{–}p_\sigma)$ bonding between the phosphorus and equatorial fluorine atoms. From each of these a 2p node is directed toward the central phosphorus atom, which exhibits an interacting ring of charge attributable to the pair of P 3p atomic orbitals having lobes lying in this plane. In addition, one of these phosphorus 3p lobes forms a $(p_\pi\text{–}p_\pi)$ bond with a suitably oriented single 2p orbital of each of the axial fluorine atoms. Orbital $4a''$ represents a reverse of the bonding found in the pair of $4e'$ orbitals in that there is F–P $(p_\pi\text{–}p_\pi)$ bonding between the phosphorus and each of the axial fluorine atoms. This bonding behavior of orbital $4a_2''$ corresponds to two pancake-shaped lumps of charge lying between the phosphorus and the axial fluorine atoms, above and below the equatorial plane of the molecule. This leads to the three parallel nodal planes depicted in Fig. 3.65 for orbital $4a_2''$.

The pair of degenerate orbitals, $1e''$, shows no occupation of the phosphorus in the basis set not containing d orbitals, whereas, with the inclusion of phosphorus 3d functions, the phosphorus charge distribution is dominated by the four lobes of a single d orbital lying perpendicular to the equatorial plane. For the pair of $1e''$ orbitals taken together, the bonding between the phosphorus and each equatorial fluorine atom corresponds to a single F–P $p_\pi \rightarrow d_\pi$ system, with the π nodal surface lying in the equatorial plane, whereas, between the phosphorus and the axial fluorine atoms, both of the $(p_\pi\text{–}d_\pi)$ contributions having nodal planes passing through the P–F bond axis are equally employed. The nodal planes of each of the $1e''$ orbitals primarily represent fluorine lone pairs, with some pure $p_\pi \rightarrow d_\pi$ feedback from the fluorine to the phosphorus.

The pair of $5e'$ molecular orbitals taken together is dominated by a combination of F–P $p_\pi \rightarrow d_\pi$ and $p_\sigma \rightarrow d_\sigma$ feedback, with only the equatorial fluorine atoms being involved. For this orbital pair, the nodal structure is such that each equatorial fluorine utilizes its pair of 2p orbitals exhibiting lobes in the equatorial plane so that there is a ring of charge lying in this plane and

surrounding each of these fluorine atoms. The overall bonding, which is confined to this plane, consists of both σ and π interactions of the ring of charge around each equatorial fluorine atom with the pair of phosphorus 3d orbitals lying in this plane.

Now moving on to Figs. 3.68 and 3.69, we see that for molecular orbital $8a_1'$ the bonding between the phosphorus and the axial fluorine may be characterized as being dominated by F–P $(p_\sigma\text{–}d_\sigma)$ interactions. Orbital $1a_2'$ is the only one which corresponds to pure fluorine lone-pair character, with the entire charge being distributed between the three equatorial fluorine atoms only. As shown in Fig. 3.65, this orbital corresponds to three nodal planes lying at a 120° angle to each other with respect to their intersection along the C_3 axis of the molecule.

Molecular orbital $5a_2''$ and the $6e'$ degenerate pair may be treated together. The similarity of their nodal structures is apparent in Fig. 3.65, where each is depicted as a flat nodal surface, coupled with approximately ellipsoidal nodal surfaces on both sides of this nodal plane. In each of these cases, the overlap population between any axial and any equatorial fluorine atom is antibonding. It is interesting to note the development of the nodal structure in the series of molecular orbitals, $6a_1'$, $8a_1'$, $5a_2''$, and $6e'$. Orbital $8a_2'$ follows from $6a_1'$ by the addition of an approximately circular nodal surface, whereas the $5a''$ plus $6e'$ trio of orbitals is obtained by adding a nodal plane to orbital $6a_1'$. This of course, parallels the development in the series of atomic orbitals 1s, 2s, $2p_x$, $2p_y$, and $2p_z$.

The degenerate pair of orbitals, $2e''$, the least stable of the filled PF_5 molecular orbitals, exhibits a node passing through the equatorial plane of the molecule and a pair of nodes passing through the molecular axis, as shown in Fig. 3.65. With this pair of orbitals, as in the case of orbital $1e''$, the P–F bonding wholly involves phosphorus d orbitals. The pair of $2e''$ orbitals exhibits F–P $p_\pi \rightarrow d_\pi$ bonding for all five of the fluorine atoms. In this pair of orbitals, each axial fluorine atom interacts with the phosphorus through a ring of charge centered on the fluorine, because both of the F $2p_\pi$ orbitals are equally filled. In contrast,

Fig. 3.68. Cross-sectional electron-density plots for the least stable valence-shell orbitals of phosphorus pentafluoride, as measured in the equatorial plane.

Fig. 3.69. Cross-sectional electron-density plots for the least stable valence-shell orbitals of phosphorus pentafluoride, as measured in the axial plane.

Fig. 3.68 (left).

Fig. 3.69 (right).

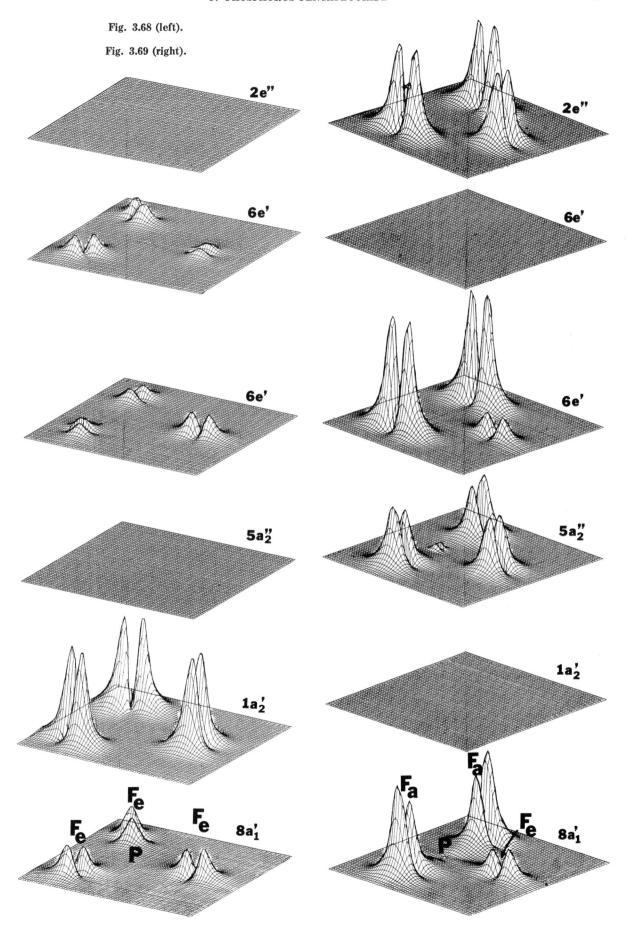

bonding of the equatorial fluorine atoms with the phosphorus involves only a single F $2p_\pi$ orbital.

In summary, note that the fluorine 2p lone-pair character is contributed to the molecule by all of the orbitals except for the five more stable ones shown in Figs. 3.63 and 3.64. The filled valence-shell molecular orbitals of PF_5 naturally fall into three classes. The first class (orbitals $5a_1'$, $3e'$, $3a_2''$, and $6a_1'$) is composed of modest amounts of low-lying phosphorus atomic orbitals interacting with the fluorine $2s_\sigma$ orbitals to give what might be described as highly polarized $(P-F)\sigma$ bonds. The second set of five orbitals ($7a_1'$, $4e'$, $4a_2''$, and $8a_1'$) is derived from the fluorine $2p_\sigma$ orbitals bonding to the phosphorus, with some P–F anti-bonding caused by the incorporation of fluorine 2s character. This mixing results in polarization away from the phosphorus and thus produces the lone pairs lying along the bond axes, n_σ. The molecular orbitals in the last class ($1e''$, $5e'$, $1a_2'$, $5a_2''$, $6e'$, and $2e''$) are dominated by the fluorine lone pairs, n_π, which are perpendicular to the P–F bond axes.

P. Internal Rotation in Molecules, with Emphasis on Diphosphine

As indicated by the fact that rotational barriers can be calculated to within a few tenths of a kilocalorie per mole using only moderately sized basis sets without polarizing functions, the SCF approximation gives a good representation of the effects on the electronic redistribution due to internal rotation in molecules. An interesting molecule for such investigations is diphosphine, H_2PPH_2, which not only exhibits twisting about a bond connecting two atoms of the third row of the periodic table but also illustrates the pronounced effect of unshared pairs of electrons in establishing the barrier to rotation. However before turning to an orbital-by-orbital discussion of internal rotation in diphosphine, we shall consider the effect of internal rotation on the total electron-density distributions of three molecules, formic acid, monomethyl phosphine, and methylenephosphorane. Electron-density distributions of the individual orbitals of a single conformation of the latter two of these molecules are presented, respectively, in Fig. 3.27 and in Figs. 3.28 plus 3.29.

HCOOH—The formate ion consists of a central carbon atom bonded to two oxygens and one hydrogen atom, with all of the atoms lying in the same plane. Formic acid exhibits a hydrogen bonded to one of the oxygen atoms at a COH

angle of 105°19′, as determined from microwave spectroscopy, with internal rotation occuring around the C–OH bond axis. The most stable conformation (the cis form) corresponds to the hydroxyl hydrogen being in the plane of the other atoms and directed toward the other oxygen atom. The next most stable form is the trans form, in which the hydrogen is again in the plane of the rest of the atoms but on the opposite side from the other oxygen. In the least stable conformation (the intermediate form) the hydroxyl hydrogen lies at right angles to the molecular plane. In the following discussion of this molecule, only the planar forms (i.e., the cis and trans) will be treated in order to simplify the discussion.

Figure 3.70 shows electron-density plots made in the plane of the molecule, as calculated [2] from a (52/52/3) Gaussian basis set. The top diagram in this figure shows the total electron density for the trans conformation and the total electron density for the cis transformation appears immediately below. The bottom diagram in Fig. 3.70 is a difference plot showing what happens to the total electron density of the formic acid when going from the cis to the trans conformation. Note that the electron-density scale in the third diagram of this figure has been made five times more sensitive than the electron-density scale of the upper two diagrams in order to emphasize the differences resulting from the rotation.

In the righthand front part of the difference plot there is a big hole, whereas in the righthand rear part there is a peak of equal size. This is due to the transfer of the hydrogen from the front to the rear of its oxygen atom and is an obvious feature of this electron-density plot. However, note that there is also a high peak of oval cross-sectional density situated on the oxygen to which the rotated hydrogen is bonded. This means that the charge on this oxygen is greater in the trans than the cis conformation. However, this is not all that has happened. In addition there are changes in the electron density between the two rotamers in the region of the carbon nucleus and the nucleus of the other oxygen atom. Careful inspection of the electron-density plot also shows a small hump in the vicinity of the hydrogen atom bonded to the carbon. Thus, in the trans form there is more charge on each of the nuclei, whereas in the cis form there is more charge in the bonding regions. Obviously therefore, the cis form should be more stable than the trans, as is the case.

We have seen that for the HCOOH molecule the detailed electron density throughout the entire molecule is affected by the rotation about the carbon–oxygen bond. This behavior, in which the entire electron density of a moderately sized polyatomic molecule is affected by the rotation of one of its parts, appears to be a very general

Fig. 3.70. Cross-sectional electron-density plots for all of the electrons in the trans and cis configurations of formic acid, measured in the molecular plane. The bottom plot in this figure corresponds to the difference in electron density between the cis and the trans configuration, with the electron-density scale being magnified fivefold as compared to the two upper plots.

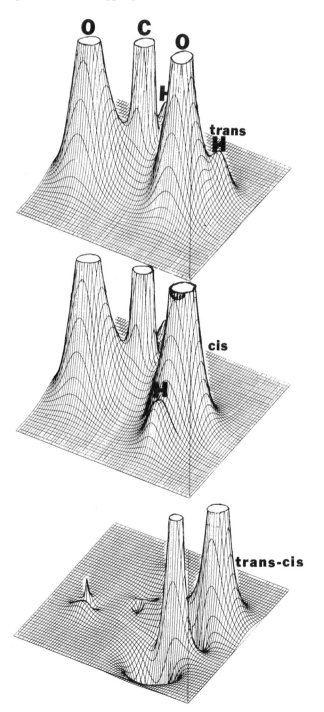

phenomenon, as evidenced by the examples of methyl phosphine (see Fig. 3.71) and methylene-phosphorane (see Fig. 3.72). Again the difference plots at the bottom of Figs. 3.71 and 3.72 have an electron-density scale magnified fivefold over that of the corresponding total electron-density plots.

CH_3PH_2—In going from the lone-pair eclipsed (middle diagram of Fig. 3.71) to the staggered form (top diagram) of methylphosphine [7], there is obviously a big change in the lefthand side of the difference plot (bottom diagram of Fig. 3.71) due to displacement of the hydrogens. But, in addition to this, there are changes in the

Fig. 3.71. Cross-sectional electron-density plots for methyl phosphine, as measured in the plane passing through the phosphorus, carbon, and one of the hydrogen atoms. The lower plot corresponds to the staggered minus the eclipsed form, with the electron-density scale being magnified five times over that of the plots above.

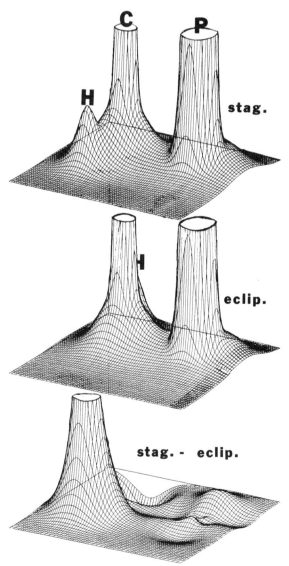

neighborhood of the phosphorus atom. Thus, from the difference plot, we see that the staggered form exhibits a greater electron-density along the C–P bond axis, as well as a pronounced increase of electron density in the region of the phosphorus lone-pair electrons, as compared to the eclipsed form. Also, the staggered form shows less electron density between the two hydrogens bonded to the phosphorus than does the eclipsed form. In short, the staggered form has more charge density in the bonding region and therefore should be more stable than the eclipsed form, as was found from the calculated SCF energies and from microwave spectroscopy.

CH_2PH_3—The top three diagrams of Fig. 3.72 correspond to the total electron density of (A) the staggered form of methylenephosphorane [8] in the plane perpendicular to that of the methylene group; (B) the staggered form in the plane including the methylene group; and (C) the eclipsed form in this latter plane. Thus, by taking appropriate differences between these three electron-density plots we can show the effect on this molecule of rotating the methylene group while holding the phosphino group fixed, or of rotating the phosphino group while keeping the methylene group fixed. The uppermost difference plot in Fig. 3.72 (the second diagram from the bottom) was obtained by subtracting total-density plot C from plot A and this plot corresponds to the rotation of the two hydrogens of the methylene group out of the plane of these plots. This rotation readily accounts for the two large holes in the right side of the difference plot. However, it should be noted that the electron density on the hydrogens bonded to the phosphorus is greater in the staggered than in the eclipsed conformation and that there is also a concomitant increase in the contributions of the phosphorus 3p atomic orbitals exhibiting lobes in the plane of the electron-density diagram at right angles to the P–C bond axis. The broad positive humps lying just behind the holes left by the two hydrogen atoms are associated with the similarly directed carbon 2p atomic orbitals. Indeed, these two humps in the difference plot are primarily attributable to the rotation of

Fig. 3.72. Cross-sectional electron-density diagrams of (A) the staggered form of methylenephosphorane in the plane perpendicular to the methylene group, (B) the same form in the plane including the methylene group, and (C) the eclipsed form in the latter plane, as well as difference plots showing effect of the rotation of CH_2 group and the PH_3 group. The bottom two plots have the electron-density scaled fivefold greater than for plots A–C.

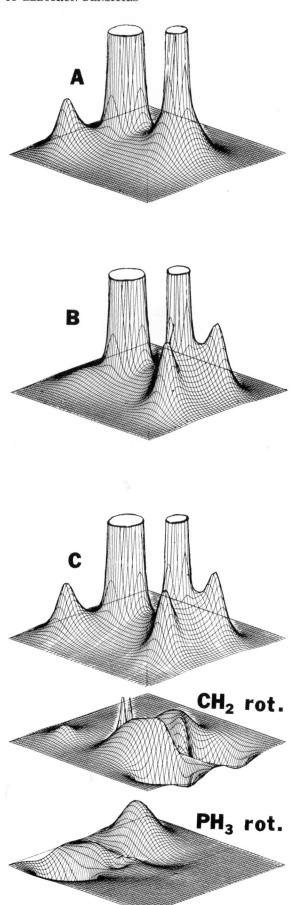

the nodal planes of the molecular orbitals shown as 2a″ and 3a″ in Figs. 3.28 and 3.29 from a position coincidental with the plane of the plot to a perpendicular position.

The bottom diagram of Fig. 3.72 is a difference plot corresponding to rotation of the PH_3 group so that the hydrogen originally at the lefthand corner of the basal plane of the plot is taken out of the plane, leaving no phosphino hydrogens in the plane. This difference plot was obtained by subtracting the total-electron-density plot C from plot B. The important thing to be seen from the bottom plot of Fig. 3.72 is that rotation of the phosphino group has essentially no effect on the methylene group of the methylenephosphorane molecule. Note in this lower electron-density difference plot that the positive hump at the rear of the diagram is attributable to the closer position of one of the phosphino hydrogens to the molecular plane, although this hydrogen does not lie in the plane. Rotation in the molecular orbital shown as 10a′ in Figs. 3.28 and 3.29 is primarily responsible for this effect.

It is important to note that the shifts in electron density observed for the methylenephosphorane molecule are mainly attributable to the four molecular orbitals that are π type with respect to the P–C bond (the axis of rotation). Reflection on the matter of internal rotation would lead one to conclude that molecular orbitals which have σ character with respect to the bond being rotated ought to have little effect on the spatial redistribution of electrons not directly associated with the motion of atomic nuclei, whereas those molecular orbitals which are π with respect to the bond axis of rotation should lead to electronic redistributions in addition to those gross shifts in spatial charge distributions that must accompany a change in nuclear positions.

The barrier to internal rotation in methylenephosphorane is extremely small, as expected for sixfold symmetry, being calculated to be 0.003 kcal/mole with d character included [i.e., a (951/52/3) Gaussian basis set] and 0.016 kcal/mole with no d character being allowed [i.e., a (95/52/3) basis set]. In the common electron-bond language of chemistry, the low barrier for internal rotation in methylenephosphorane indicates that there is no double-bond character in the P–C bond. However, in this language, the bond could also be considered as involving resonance between a single-bond and a triple-bond structure, with the latter (if present) being associated with concomitant P–H and C–H no-bond resonance contributions. Although twisting a double bond involves much energy, neither a single nor triple

bond should in themselves exhibit any hindrance to rotation. It is interesting to note that, although the rotational barrier for methylenephosphrane is inappreciably small, the relative energies of some of the orbitals are interchanged by rotation. Thus, for the staggered form, a listing of the valence orbitals in order of increasing energy is 6a′, 7a′, 2a″, 8a′, 9a′, 3a″, and 10a′, whereas for the eclipsed form the ordering is 6a′, 7a′, 8a′, 2a″, 9a′, 10a′, and 3a″.

P_2H_4—It is common to discuss internal rotation in terms of a relatively small number of rotational isomers. For diphosphine, these are the eclipsed, gauche, semieclipsed, and staggered forms, depicted below in a view looking down the P–P bond axis:

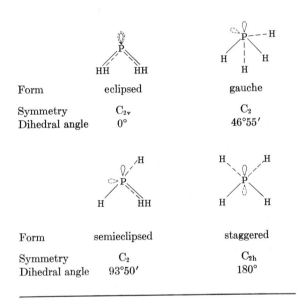

Form	eclipsed	gauche
Symmetry	C_{2v}	C_2
Dihedral angle	0°	46°55′

Form	semieclipsed	staggered
Symmetry	C_2	C_{2h}
Dihedral angle	93°50′	180°

Fig. 3.73. Cross-sectional electron-density plots of the valence-shell molecular orbitals of the eclipsed form of diphosphine, measured in a plane passing through the two phosphorus atoms and lying parallel to the line passing through the hydrogen atoms of the lefthand PH_2 group.

Fig. 3.74. Cross-sectional electron-density plots of the valence-shell molecular orbitals of the gauche form of diphosphine, measured in the plane passing through the two phosphorus atoms and lying parallel to the line passing through the hydrogen atoms of the lefthand PH_2 group.

Fig. 3.75. Cross-sectional electron-density plots of the valence-shell molecular orbitals of the semieclipsed form of diphosphine, measured in the plane passing through the two phosphorus atoms and lying parallel to the line passing through the hydrogen atoms of the lefthand PH_2 group.

Fig. 3.76. Cross-sectional electron-density plots of the valence-shell molecular orbitals of the staggered form of diphosphine, measured in the plane passing through the two phosphorus atoms and lying parallel to the line passing through the hydrogen atoms of the lefthand PH_2 group.

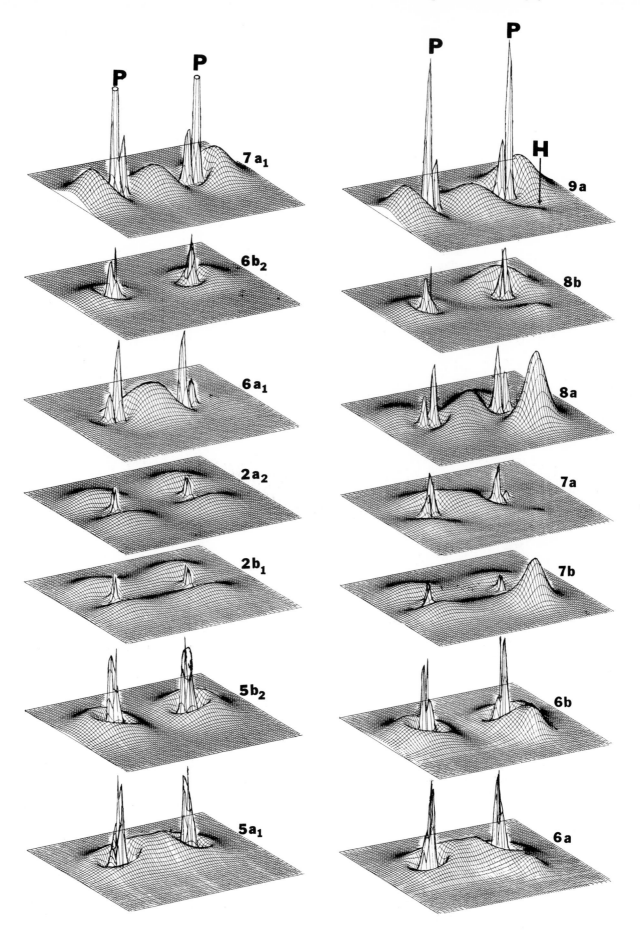

Fig. 3.73. Eclipsed P₂H₄.

Fig. 3.74. Gauche P₂H₄.

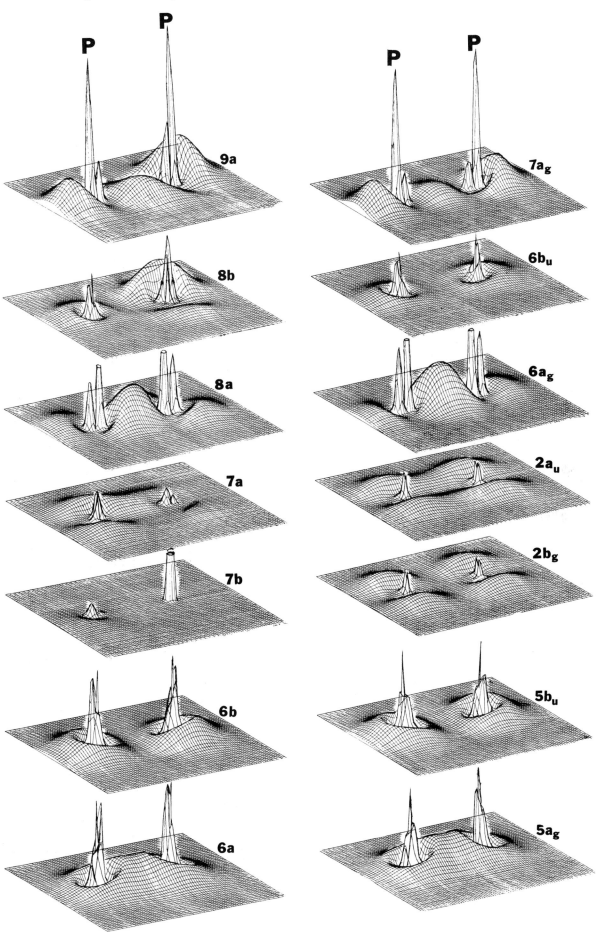

Fig. 3.75. Semieclipsed P₂H₄.

Fig. 3.76. Staggered P₂H₄.

The calculations [17] on both hydrazine [in a (93/3) Gaussian basis set] and diphosphine [in an (84/2) basis set] show that there are two minima and two maxima in the energy during a 180° rotation of the dihedral angle corresponding to the transition from the eclipsed to the staggered form. The least stable conformation is the eclipsed form, and the second maximum in the rotational barrier occurs partway between the semieclipsed and the staggered form. The most stable rotamer corresponds to a configuration between the gauche and the semieclipsed and the other lesser minimum corresponds to the staggered conformation.

Figures 3.73–3.76 show the seven filled valence-shell molecular orbitals of diphosphine as observed in a plane passing through the two phosphorus atoms and parallel to the line connecting the left-hand pair of hydrogen atoms. The latter remain in a fixed position, whereas the hydrogen atoms at the righthand side of each diagram are rotated to transform from one conformation to another. In these figures, the valence-shell molecular orbitals are shown in decreasing order of stability (reading from the bottom up) only for the gauche form. For the other conformations some alteration in the ordering with respect to energy has been introduced in order to show the related orbitals of the different rotamers on the same line. The reader is referred to Fig. 3.77 for an energy-level diagram in which the orbitals that exhibit related

Fig. 3.77. An orbital-energy plot for the various rotamers of diphosphine, showing the related molecular orbitals discussed in the text.

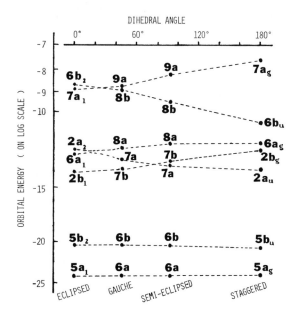

electron-density distributions are connected by dotted lines. In the following interpretation of the effect of rotation on the molecular orbitals of diphosphine (as shown in Figs. 3.73–3.76 and 3.78–3.81), the name of the molecular orbital of the gauche conformation will be the key reference; but it is to be understood that the same comments apply to the related orbital of each of the other conformations. Note that the scale of these figures was chosen so that the electron density is magnified four times that usually employed in this book so that the changes in the diffuse outer antinodes of the phosphorus atomic orbitals become readily apparent.

Molecular orbital 6a of the gauche form involves overlap between the s-type valence orbitals of all of the atoms. As expected, this orbital is invariant to rotation except for the situation (e.g., the gauche form of Fig. 3.74) in which one or the other of the pair of rotating hydrogen atoms comes close to the cross-sectional plane in which the electron density is being represented. Gauche-form orbital 6b also involves only valence-shell s atomic orbitals but differs from molecular orbital 6a in exhibiting a nodal plane, which bisects the P–P bond. Similar to orbital 6a, molecular orbital 6b should also be unchanged by rotation. Gauche-form orbital 7b exhibits P–P π bonding (i.e., π bonding with respect to the axis of rotation) as well as P–H (p_σ–s_σ) bonding between each phosphorus and its pair of contiguous hydrogen atoms. The presence of the P–H σ bond in this particular orbital establishes a fixed polarity for the phosphorus 3p lobes involved in the concurrent P–P π bond. Note in Figs. 3.73–3.76 that orbital $2b_1$ of the eclipsed form shows p_π–p_π bonding (rather than antibonding), which means that the lobe polarity of the p_π atomic orbitals of the two phosphorus atoms is in the same direction. This is still approximately true for the related orbital 7b of the gauche form, but for orbital 7b of the semieclipsed form the rotation of the hydrogen atoms has caused these 3p atomic orbitals on the pair of phosphorus atoms to be crossed with respect to each other, so that the P–P π bonding disappears. Finally, for orbital $2b_g$ of the staggered form, these p_π orbitals on the pair of phosphorus atoms have been completely flipped over so that the P–P π contribution to orbital $2b_g$ of the staggered form is antibonding.

Molecular orbitals $6a_1$ and $2a_2$ of the eclipsed form exhibit closely lying energies, as can be seen from Fig. 3.81. Because any slight rotation about the P–P bond causes these orbitals to have exactly the same symmetry designation (i.e., they are then both simply A-type orbitals), the noncrossing rule indicates that, with respect to energy, orbital 8a of the gauche form should be correlated with orbital $2a_2$ of the eclipsed form and 7a of the gauche with $6a_1$ of the eclipsed, as shown in Fig. 3.77. However, when orbital mixing is not extensive, it is usually found that the electronic distribution does exhibit crossing; the series of plots in Figs. 3.73–3.76 starting with orbitals $6a_1$ and $2a_2$ of the eclipsed form are seen to show this behavior.

Orbital $2a_2$ of the eclipsed form corresponds to P–P $(p_\pi - p_\pi)$ antibonding, whereas orbital $6a_1$ corresponds to P–P $(p_\sigma - p_\sigma)$ bonding. In contrast, molecular orbitals 7a and 8a of the gauche form both have the phosphorus 3p atomic orbitals directed at an angle to the P–P bond axis instead of perpendicular and parallel to it as was necessarily the case for orbitals $2a_2$ and $6a_1$ of the eclipsed form. The rotation away from the P–P bond axis of the phosphorus 3p lobes in orbital 7a of the gauche form is opposite in sign from that of orbital 8a of this form. Thus, it appears that the 3p contributions of the phosphorus atoms are turning with the internal rotation in order to preserve the bonding between each phosphorus and its pair of hydrogens as well as between the two phosphorus atoms during rotation about the P–P bond. Continuing the rotation leads to orbitals 7a of the semieclipsed form and $2a_u$ of the staggered form, which correlate well with orbital $2a_2$ of the eclipsed form. Again, we find that orbital $2a_2$ of the eclipsed form is P–P π antibonding, whereas orbital $2a_u$ of the staggered form is P–P π bonding, in accord with the arguments given for the previously described orbitals ($2b_1$ of the eclipsed form and $2b_g$ of the staggered). Orbitals 8a of the semieclipsed and $6a_g$ of the staggered exhibit P–P $(p_\sigma - p_\sigma)$ bonding, just as was found for orbital $6a_1$ of the eclipsed form.

Orbitals 8b and 9a of the gauche form are dominated by the phosphorus lone pair with some P–P σ bonding in orbital 9a. As might be expected, the electron densities of these orbitals are little changed by the rotation, except for the angular position of the righthand phosphorus electron pair, which must turn with the rotation of this PH_2 group. This PH_2 rotation also leads to an angular distortion of the electron density in the P–P bonding region of orbitals 9a of the gauche and semieclipsed forms.

Figures 3.78–3.81 are similar to Figs. 3.73–3.76, except that the cross-sectional plane in which the electron density is plotted passes again through the P–P bond axis but is at right angles to the plane of Figs. 3.73–3.76. In other words, in Figs. 3.78–3.81 the basal plane passes through the P–P bond axis and bisects the line connecting the pair of hydrogen atoms on that phosphorus atom which appears at the lefthand side of each diagram. Note that the bottom two rows of diagrams in Figs. 3.78–3.81 are quite similar to the bottom two rows in Figs. 3.73–3.76, as would be expected from the fact that these molecular orbitals involve the interaction of the spherically symmetrical valence-shell s atomic orbitals and hence, except for the positioning of the hydrogen atoms, will look about the same in any cross-sectional cut passing through the P–P bond axis. Orbitals $2b_1$ and $2a_1$ of the eclipsed form and $2b_g$ and $2a_u$ of the staggered form exhibit no electron density in the plane chosen for Figs. 3.78 and 3.81, in accord with the P–P π character that they show in Figs. 3.73 and 3.76. For orbital 7b in Figs. 3.79 and 3.80, the interaction with the hydrogen atoms shows up, along with some phosphorus lone-pair character in the latter figure.

It was previously noted that orbital 7a and 8a of the gauche form resulted from mixing between orbitals $2a_2$ and $6a_1$ of the eclipsed form. This effect may be seen in Figs. 3.78 and 3.79, where the contribution of orbital $6a_1$ of the eclipsed form to 7a of the gauche form is particularly apparent. Figure 3.80 shows the continuation of this effect for the semieclipsed form. Comparison of the

Fig. 3.78. Cross-sectional electron-density plots of the valence-shell molecular orbitals of the eclipsed form of diphosphine, in a plane perpendicular to the plane of Fig. 3.73.

Fig. 3.79. Cross-sectional electron-density plots of the valence-shell molecular orbitals of the gauche form of diphosphine, in a plane perpendicular to the plane of Fig. 3.74.

Fig. 3.80. Cross-sectional electron-density plots of the valence-shell molecular orbitals of the semieclipsed form of diphosphine, in a plane perpendicular to the plane of Fig. 3.75.

Fig. 3.81. Cross-sectional electron-density plots of the valence-shell molecular orbitals of the staggered form of diphosphine, in a plane perpendicular to the plane of Fig. 3.76.

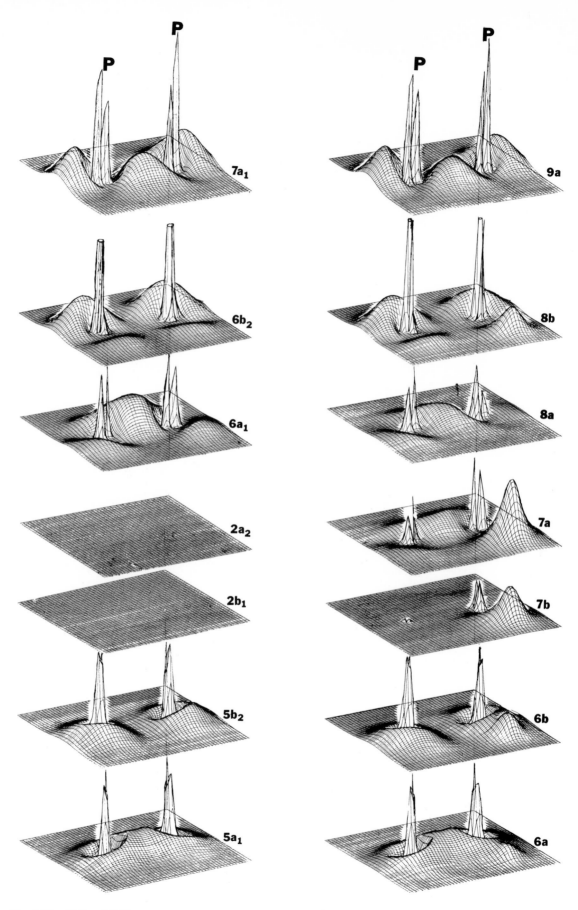

Fig. 3.78.   Eclipsed P_2H_4.

Fig. 3.79.   Gauche P_2H_4.

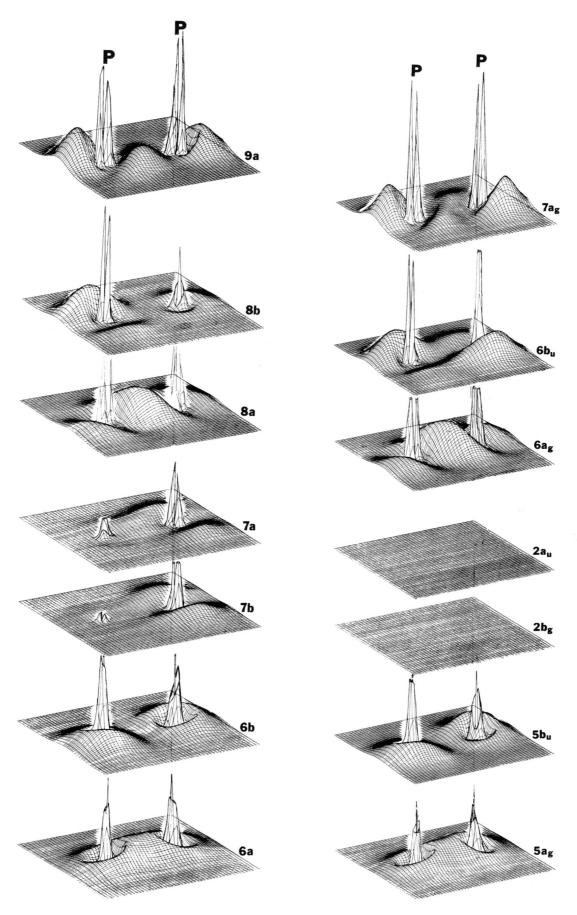

P   **P**

9a

8b

8a

7a

7b

6b

6a

P   **P**

7a$_g$

6b$_u$

6a$_g$

2a$_u$

2b$_g$

5b$_u$

5a$_g$

Fig. 3.80. Semieclipsed P$_2$H$_4$.

Fig. 3.81. Staggered P$_2$H$_4$.

97

cross-sectional plot of the eclipsed form of orbital $6a_1$ in Fig. 3.78 with the plot of the same orbital in Fig. 3.73 shows that this orbital appears symmetric when the cross section is cut perpendicular to the C_2 axis of the molecule and curved when the cross-sectional plane passes through this axis. This behavior is also clearly seen in the upper two rows of orbitals in Figs. 3.78–3.81. Note that in the plane of these figures the lone pairs of the phosphorus atoms appears on the same side of the P–P bond axis for the eclipsed form and on opposite sides for the staggered form, as would naturally be expected.

Comments

It has often been remarked that, as more theoretical effort is directed to the study of internal rotation of molecules, the more complicated and less readily comprehensible the process appears. However, we believe that the use of electron-density plots helps greatly to clarify and illuminate this problem, as indicated by the above examples.

From the electron-density plots, it appears that there is physical twisting of the electron cloud in the P–P bonding region of P_2H_4 for the two molecular orbital sequences starting with orbital $6a_1$ and orbital $2a_2$ of the eclipsed form of diphosphine. This behavior is attributable to the fact that these orbitals are quite delocalized (as shown by the Mulliken population analyses) so that there is considerable electronic charge shared between each phosphorus and its neighboring hydrogen atoms as well as between the phosphorus atoms. Because all of this charge is interconnected, moving one pair of hydrogen nuclei with respect to the other will indeed lead to an obvious angular distortion (i.e., twisting) of a charge distribution that is in the region of the P–P bond axis but is not cylindrically symmetrical with respect to it. In other words, there is a strong tendency to preserve the delocalized character of these molecular orbitals during the course of rotation about the P–P bond. For the sequence of orbitals starting with $5a_1$ of the eclipsed P_2H_4 molecule, no such twisting of the P–P bond is observed, because these orbitals exhibit cylindrical symmetry about the P–P bond axis.

REFERENCES

1. W. E. Palke and W. N. Lipscomb, *J. Amer. Chem. Soc.* **88**, 2384 (1966).
2. I. Absar and J. R. Van Wazer, unpublished results.
3. A. D. McLean, *J. Chem. Phys.* **39**, 2653 (1963).
4. J. B. Robert, H. Marsmann, I. Absar, and J. R. Van Wazer, *J. Amer. Chem. Soc.* **93**, 3320 (1971).
5. C. S. Ewig, J. M. Howell, and I. Absar, unpublished results; J. B. Robert, H. Marsmann, L. J. Schaad, and J. R. Van Wazer, *Phosphorus* **2**, 11 (1972).
6. E. Clementi, D. L. Raimondi, and W. P. Reinhardt, *J. Chem. Phys.* **47**, 1300 (1967).
7. I. Absar and J. R. Van Wazer, *J. Chem. Phys.* **56**, 1284 (1972); I. Absar and J. R. Van Wazer, *Chem. Commun.* p. 611 (1971).
8. I. Absar and J. R. Van Wazer, *J. Amer. Chem. Soc.* **94**, 2382 (1972).
9. H. Marsmann, J. B. Robert, and J. R. Van Wazer, *Tetrahedron* **27**, 4377 (1971).
10. I. Absar, L. J. Schaad, and J. R. Van Wazer, *Theor. Chim. Acta* **29**, 173 (1973).
11. J. R. Van Wazer and I. Absar, *Adv. Chem. Ser.* **110**, 20 (1972).
12. I. Absar and J. R. Van Wazer, *J. Phys. Chem.* **75**, 1360 (1971).
13. I. Absar and J. R. Van Wazer, *J. Amer. Chem. Soc.* **94**, 6294 (1972).
14. J. M. Howell and J. R. Van Wazer, *Inorg. Chem.* **13**, 737 (1974).
15. A. Strich and A. Veillard, *J. Amer. Chem. Soc.* **95**, 5574 (1973).
16. J. M. Howell, J. R. Van Wazer, and A. R. Rossi, *Inorg. Chem.* **13**, 1747 (1974).
17. I. Absar, J. B. Robert, and J. R. Van Wazer, *J. Chem. Soc., Faraday Trans. 2* **68**, 1055 (1972); J. B. Robert, H. Marsmann, and J. R. Van Wazer, *Chem. Commun.* p. 356 (1970).

Index

Physical Chemistry

A Series of Monographs

Editor: **Ernest M. Loebl**

Department of Chemistry

Polytechnic Institute of New York

Brooklyn, New York

A 5
B 6
C 7
D 8
E 9
F 0
G 1
H 2
I 3
J 4